Planet Pragmatism: The New Path to Prosperity

a playbook of common sense for the common good

Mark C. Coleman

Wildebeest Publishing Company, LLC
Syracuse, NY

Do you have a story to tell? What's your animal spirit? Share it with us. #hellobeesties

Copyright © 2025 by Mark C. Coleman

You may visit the author's website at www.markcolemaninsights.com

Wildebeest Publishing Company, LLC

All rights reserved, including the right to reproduce this book or portions thereof in any form. Unauthorized copying or distribution of the book is forbidden.

For more information about copyrights and usage, special discounts on bulk purchases, workshops, and engagements, please contact Wildebeest Publishing Company, LLC at (315) 220-0217, info@wildebeestpublishing.com, or online at www.wildebeestpublishing.com
Wildebeest Publishing is dedicated to providing flexible remote work opportunities and has a presence in Syracuse, New York City, Tampa, and Denver.

Wildebeest Publishing Company, LLC paperback First Edition June 2025, United States of America

ISBN 978-1-958233-43-6 (paperback)
ISBN 978-1-958233-44-3 (ebook)
LCCN 2025911785

The publisher and artists can accept no legal responsibility for any consequences arising from the application of information, advice, or instructions given in this publication. The author, artists, and publisher have made all reasonable efforts to contact copyright holders for permission and apologize for any omissions or errors pertaining to credit for existing works. Corrections may be made to future versions. The opinions expressed in the book are those of the author and do not necessarily reflect the views of the publisher or the foreword author.

PRAISE FOR PLANET PRAGMATISM

"In Planet Pragmatism, Mark Coleman offers a timely invitation to reimagine life and leadership in service to our common good. Drawing on decades of professional experience, Coleman blends personal narrative, social commentary, and systems thinking to guide us beyond polarization and into a principled approach that serves people and planet. I recommend this distinctive text for anyone seeking clarity in response to crisis."

– Rev. Brian E. Konkol, Ph.D.
Vice President and Dean of Hendricks Chapel and
Professor of Practice, Maxwell School of
Citizenship and Public Affairs
Syracuse University

"In this global game, the smartest move is the one that uplifts all players. Pragmatic climate action must be informed by local truths, regional priorities, and the universal need for dignity, prosperity, and a liveable planet."

– Princy Mthombeni
Founder, Africa4Nuclear

"In Planet Pragmatism, Mark brings a unique, pragmatic and multi-dimensional approach to sustainability as evidenced in his frameworks. Mark showcases his innate ability to dissect a complex topic from a multi-disciplinary lens, providing us with a better understanding of the interconnectivities across all the factors driving sustainability and delves deeper into a thought-provoking analysis asking leaders to self-reflect upon their own decision-making

abilities to catalyze change. Planet Pragmatism bridges intellectual depth with practical, real-world insights. This is a must-read for any sustainability leader who understands the multi-dimensional challenges of driving lasting change and offers a powerful lesson in leadership for navigating executive decision-making in a constantly evolving landscape."

– Nidhi Chadda
Founder and CEO, Enzo Advisors

"Mark C. Coleman's Planet Pragmatism is a comprehensive exploration of how individuals, institutions, and societies can recalibrate their pursuit of prosperity in a rapidly changing world. The book offers a framework called "Planet Pragmatism" rooted in practical wisdom, ethical foresight, and common-sense approaches for addressing the world's challenges—ranging from climate change to societal polarization, economic inequality, and the ethical use of technology like AI. In the spirit of pragmatism Mark shares tools to help organizations make wiser choices for the planet and society."

– Al Iannuzzi, Ph.D.
Author of Greener Products: The Making and Marketing of Sustainable Brands

"Planet Pragmatism" is a masterfully distilled compilation of all those ingredients needed to achieve prosperity in the 21st century, under the condition of sustainable results. Yes, it is a playbook that convincingly progresses from fact-based lessons learned towards a well-defined path of action. After all, pragmatism works only when it is firmly founded on a principled, substance-rich framework. What I see in plain terms is a recipe for a future where moral values do not necessarily contradict corporate strategy and individuality does not surrender under the pressure of a vaguely defined common good. A very modern approach to a quest as old as the history of human reasoning. This is a vibrant and most relevant wake-up call as sustainability may not be dead (to paraphrase a chapter title), yet everything points to a meta-sustainability era coming, very much in need of pragmatic solutions.

– Stelios Vogiatzis
Director, Consulting Services, Forvis Mazars Greece

"In a world filled with uncertainty, if you are someone who has chosen a purposeful path, guided by principles, while navigating systemic barriers and striving to inspire the next generation to live with intention, then this book is for you. The author provides hope in how collective intelligence, community-driven approaches, and holistic wellness, using pragmatic frameworks and tools to support human wisdom, can lead us towards Planet Prosperity. More than just an insightful read, this book serves as a practical playbook for the everyday citizen, helping us navigate the complexities of today's world with resilience and purpose. Let's lead with hope and humanity, crafting solutions that light the way forward."

– Parvathy Thanumoorthy
Business Transformation Leader | Community Wellness Advocate

"In Planet Pragmatism, *Mark Coleman's words remind us that common sense, compassion, and leadership grounded in values can guide us through uncertain times. This book is both a wake-up call and a guide, showing how principled leadership and purposeful action can help us move forward with courage and care."*

– Karina Cheng
CEO of Kermise Inc., and Former Student

"The future isn't NIMBY—it's YES in my backyard. When communities recognize that land, energy, and infrastructure can be designed as regenerative assets — not liabilities — they don't resist change, they lead it.

That's why Planet Pragmatism is a powerful and timely read. Mark Coleman isn't just naming our challenges—he's offering a principled roadmap toward decentralized, dignified, and distributed prosperity.

As a decentralist and edge economist, I see this book as a critical bridge between today's broken systems and tomorrow's agentic, regenerative communities. If you're ready to move from polarization to pragmatism, from resistance to resilience—read this book."

– Pamela Norton
Founder, *TitleChain Foundation* | Architect of the M5 Economy Creator, *Minera Index & Exchange* – Sovereign Digital Asset Markets

"Mark has clearly communicated the critical elements for pragmatic prosperity for our planet in this, his fourth book! As we have discussed for many years, the convergence of complex environmental problems today and success are best served by objective adaptation, pragmatism, moderation and common-sense decisive leadership. I am honored to be considered one of Mark's mentors in his journey."

– Steve Myers
Founding Member and President, Myers Environmental Consulting, LLC

"Planet Pragmatism very neatly ties together all the simple things we need to understand about human living and sustainability through looking at ourselves. Mark Coleman skillfully weaves together the most palpable elements of the human equation to show it all leads to the freedom to pursue our own collective fate. While productive sustainability can lead to promising prosperity, Mr. Coleman aptly reveals how humans must change not only to survive but to reverse our own self-inflictive harms. This is a must-read book to live the vision of a new and promising future!"

– Eric McLamb
Founder, Chairman & CEO
Ecology Prime Inc.

"A masterful exploration of pragmatic thinking in an age of uncertainty, Planet Pragmatism challenges readers to embrace adaptable, real-world solutions. Coleman's insights illuminate a path toward clarity and action in even the most complex situations."

– Jason Torreano
CEO, Inkululeko

"In his 4th book, Mark Coleman again demonstrates his passion to see a more just, sustainable, and pragmatic world for all. Another first-class read, by a first-class friend, mentor, professor, and human being."

– Ibrahim Tahir
Former Student and Clean Transportation Specialist

"This book is dedicated to the incredibly talented sustainable enterprise and management students that I have had the pleasure of working with. With admiration and joy, I have witnessed your development as a new generation of young professionals dedicated to creating a better world. Along the way, you have challenged and expanded my understanding of our diverse and abundant planet and have enriched my personal and professional growth and development. For this, I am forever grateful. May you continue to learn, lead, and grow with the creativity, courage, confidence, clarity, curiosity, and compassion that is necessary in our changing times, and which enable your pursuit of prosperity through principles of planet pragmatism."

With deepest gratitude,
Mark Coleman

CONTENTS

PRELUDE

Feeding Faith in the Furor of Fog

A Poem Inspired by Rev. Brian E. Konkol, Ph.D., Vice President and Dean of Hendricks Chapel; Professor of Practice, Maxwell School of Citizenship and Public Affairs, Syracuse University

A thick fog blankets over life from time to time,
Clouding life's perennial sunrises and sunsets.

Deep roots provide the foundation to reach beyond the fog,
So that life-nourishing light converts to energy and provides strength,
Illuminating a path of purpose, prompting personal growth.

But then, there are seasons of life,
Where even the most prepared find themselves ill-equipped,
When the rise or our reach fails to meet the light,
As gracefully as it once did.

Engulfed by a relentless fog,
Once clear paths now are,
Unseen and out of reach,
And the mind roils in a deep sea of uncertainty.

The light of enjoyment, satisfaction, and purpose fade into a hazy abyss,
As a shroud of fear and anxiety of the unknown consume,
Then, contort conscious perception and immobilize action.

Grounded by deep roots and extended by an affirming reach,
The smothering fog gives way to a conviction of knowing.

For fog is nothing more,
Then droplets of water suspended in air,
The basic elements of life itself.
It is the fog, much like the light,
That provides needed nourishment to the roots,
Opening capillaries of connection which deliver,
An innate wisdom and eternal hope,
Ever extending the reach outward and upward,
Until the canopy collides once again,
With the mercy of brilliant light.

As the crown shimmers, a rejuvenating energy flows,
Reviving and strengthening the roots below.
And the path is renewed.

Resolve lifts the cloud of panic,
Away to reveal,
That the path itself has not changed.
It remains steadfast,
As it was once before, and has since been,
But made even more devout by the understanding,
That abstractions of life's passing seasons provide,
The faith and fervor to continually illuminate the honourable path,
Especially when it cannot be seen.

For the path, just like the light,
Has always been and shall remain accessible.

Fog and fear are temporary,
Yet essential to further extended roots and enrich reach,
Beckoning and providing courage to the crown,

To rise and shine, always,
So that we may continue to travel,
A path of righteous resolve.

Reconnecting with a Friend to Interpret the Fog of Life

Recently, I had the pleasure of lunching with a dear friend and colleague, the *Rev. Brian E. Konkol, Ph.D.*, Vice President and Dean of Hendricks Chapel and Professor of Practice, Maxwell School of Citizenship and Public Affairs, Syracuse University.

I met Dr. Konkol for lunch on the Syracuse University campus. We grabbed a bite at the wonderfully renovated Schine Student Center, and then sat down at his office in Hendricks Chapel for an informal conversation. We covered a great deal of ground in the hour we shared. We spoke about a wide range of topics spanning our families and careers, the promise and existential threat of artificial intelligence, and the philosophical underpinnings of the human condition as viewed through the lens of religious ideologies.

Dr. Konkol provides leadership at Hendricks Chapel, which is the spiritual beating heart of the Syracuse University campus and extended community. Hedricks Chapel is student-centered and a global home for religious, spiritual, moral, and ethical life. Guided by an unwavering mission and adept leadership, the Chapel serves as a hub for holistic life, preparing engaged citizens, scholars, and leaders for participation in a changing global society. Hendricks Chapel serves many incredible purposes. On a personal level, I have found it to be a place of connection and understanding, compassion and care, discovery and growth, and much more. Given the political and societal changes that are underway in the U.S. and around the world, the incredible people and work behind Hendricks Chapel are needed more than ever.

If you've never had the distinct pleasure of meeting Dr. Konkol, trust me when I say he has an uncanny ability to evoke trust, confidentiality, truth, empathy, and joy. He personifies the "judgment-free" zone, in that he represents one of those unique individuals you find yourself sharing freely with, because the conversation feels, and truly is, "safe." I met with Dr. Konkol to reconnect with a colleague, but soon found myself as a patient and patron, benefiting from the instinctive wisdom he bestows, which can only be attained from years of dedicated experience serving as a devout spiritual leader, internationally, and within the Syracuse community.

Our conversation shifted to the recent Presidential election, and Dr. Konkol shared a metaphor that stirred this personal reflection and inspired the poem at the beginning of this preface. I asked Dr. Konkol if he has seen a change in the attitude or "temperature" of the campus since the U.S. presidential election. Dr. Konkol reflected that the outcome of the election is clearly important, but it is also critical in how individuals and institutions learn, grow, and lead from here. He revealed that it is convenient, if not easy, for any individual or organization to be a "thermometer," that is, a simple gauge for measuring and reporting on the temperature of the election outcome, or 'state of society'.

We all have had a reaction and opinion regarding the outcome at this point. Some people are, quite literally, hot. Others feel like they are finally at the right temperature and temperament. Hopefully, the political temperature will not continue to rise, and people will become more conditioned to the changing ambient conditions, even if it is not their favorite season (politically speaking). The political rhetoric has been supercharged for a long time, and people are burnt out on the pendulum of extremism, something that I will discuss in greater detail later in this book.

We had a good laugh for a moment. As two longstanding residents of Upstate New York, we both have experienced the relatively critical and discontented culture of Upstate residents, particularly when it comes to weather and the changing seasons. Upstate residents are staunch weather and Syracuse University sports fanatics. But ask a Central New Yorker how their day is going, and chances are they will mention the weather before last night's basketball game. "Well, it's sunny today! But it's colder than last week, and the forecast is calling for snow showers on Friday!" Catch them on a seemingly perfect, dry, warm day, and the likelihood is that it's too warm and "we could use some rain! Are you going to the game?" You get the picture; people like to be thermometers to changing conditions, meteorological, political, sports, or otherwise.

Dr. Konkol then noted that it is far more challenging, but essential, for individuals and institutions to be more like a thermostat. A thermostat is a gauge that doesn't just monitor the external temperature; it is a device that intuitively adapts and proactively changes the ambient condition. With a thermostat, the temperature is adjusted to be cooler or warmer, and programmed, depending upon the level of comfort we seek throughout the day or year. From an individual and institutional perspective, we need to be more like thermostats, which can intuitively adapt to changing temperature and modify the environment to the user's comfort level. Conversely, thermometers only provide a single static measure of the temperature.

Change is always happening. It has been said that if there is one constant, it is change. When change happens, and specifically, change that we may not agree with, we have a fundamental choice. We can sit idly by and react like a thermometer, pontificating grievances; or we can choose to dial up or down our engagement with the change in more productive and meaningful ways.

Accepting change does not mean that you necessarily agree with change. Accepting change, however, allows you to reflect upon, rejuvenate, and recalibrate your values, so that you can remain more agile and adaptive to the changing world around you. Thus, change management is as much an exercise in temperance and understanding as it is in reaction to external mediums. Dr. Konkol explained all this far more eloquently than I have summarized here. His metaphor, however, was timely and deliberate, and certainly resonated with me.

Most would agree that this past election season was tumultuous and exhausting. There is much we stand to learn about ourselves and each other in reflecting upon whether we have (and continue to be) serving as thermometers or thermostats in our daily lives. Let me say, if you ever get the itch to reach out to a friend or colleague to simply catch up, do it. And may the time you share with your colleague be as welcoming and inspiring as having a conversation with Brian Konkol.

The U.S. Election Thermometer Just Signaled Massive Social Change

The U.S. election is readily afoot in the hours following the November 5th election result. Definitive announcements on policy, cabinet picks, and immediate executive decisions gushed from the Trump transition team at a rate unseen in prior post-election cycles. Regardless of his motivation or what he learned from his first Presidential term, Trump's 2024 post-election demeanor is one of clear, decisive intent. Voters made their choice, and he is raring to make change, *quickly*.

As many Americans celebrated and cheered the election result, others stood absolutely dumbfounded and dismayed. Yet, in the days and weeks following the election, reality for those who did not vote for President Trump would begin to set in. Millions of voters would feel misguided and estranged from their personification of America. How is it that so many would misread and misunderstand the multifaceted change that millions of others so desperately want to see happen? The post-election analysis of America and its fragmented citizenry has, in just a few short weeks, already been significant. I'm confident this election result will continue to witness a high degree of analysis and interpretation for years to come.

In reflection of my conversation with Dr. Konkol, election cycles can also generally be characterized like a thermometer. Elections are a gauge by which to measure the temperament of citizens against the leadership in office. If citizens like the incumbent administration's performance, they vote in favor of keeping the incumbent in power. If citizens dislike the performance, they vote for change. This election cycle was, however, much different than the simplistic binary temperature gauge of political alignment against progress and values. The visceral undercurrent of this past election was (and remains) a ravenous call to recalibrate American prosperity, defined not only by an economic metric, but one that also envelops national security, conservative values, and a push for policy pragmatism, even over pluralism.

The long-term aftermath of a Trump victory is difficult to predict. A Trump Presidency 2.0 will make significant near-term policy changes pertaining to domestic security, global defense and diplomacy, the economy, education, public health and healthcare, and regulatory controls. At face value, some of the changes stirring align with the pendulum shifts associated with the political cycle. This most recent shift to populism signals a swing in American tolerance for more liberal values.

Change is dynamic. It is also interesting in that it can provide both stability and uncertainty simultaneously. The Trump victory poses to have systems-level disruptive implications for education (Universities and K-12), healthcare (pharmaceutical companies, healthcare providers, insurance carriers), infrastructure (energy, telecommunications, water, and transportation), and national security (automation and decentralization of defense, and warfare), among other institutions.

Most long-standing and large institutions are always subject to change, and subsequently, they have survived by adopting incremental changes in how they interact with politics, the market, and society. Institutions have historically adapted generally in step with the ever-changing needs of their constituents. Not everyone agrees with this sentiment, however. Post-election, the Trump transition team has amplified its messaging to radically transform the U.S. government. The sentiment is that managing change through incrementalism is the antithesis to the more expedient results that most Americans desire for recalibrating the economy, for example.

Managing Change Requires Us to Be More Like a Thermostat, Intuitive and Adaptive

Time will tell the full intent, swiftness, and severity of change that a Trump Presidency 2.0 will bring upon major institutions and the broader U.S. and global society. One thing is certain, however. Managing change will look a lot different over the next 100 days than it did four years ago when the Biden administration transitioned into office. The mindset of the U.S. populace has shifted into action. The ground game and rules for pursuing and attaining prosperity are fluid and evolving. This is, after all, democracy at work, at least for the time being.

From here on out, applied intuitive and adaptive change management is the name of the game. That means being ready for anything, anytime, and anywhere. Further, this means institutions need to adopt a preventive, predictive, and proactive posture toward the pursuit of their mission and attainment of long-term prosperity.

For some, this moment of upheaval will be an exhilarating and long-awaited time for rejoicing and renewal. For others, this may be a time of reflection and repudiation.

And yet for others, this may be a time of agonizing, abhorrent retreat. Smart institutions will acknowledge that there is no bluff to be played here. Trump has shown the "Trump card," and has had over four years building upon a populist base, all calling for a sharp realignment of American values. In doing so, incremental institutional change management has been called out and is now rendered a relic of the past.

Unwanted change can cripple and immobilize action, particularly when people and institutions are at their most vulnerable state. Change management is a state of mind. Those who manage change well are resilient in the moment. But more importantly, they are highly adept, much like Dr. Konkol, at recalibrating their mindset, mission, and management to remain adaptive to navigate ambiguity and uncertainty. In the poem, "Feeding Faith in the Furor of Fog," which opened this book, I attempt to creatively relate massive change to the disorientation of fog but also acknowledge that the fog is a temporary necessity in fueling a stronger conviction going forward. I also chose to introduce the book with this inspired poem because, like faith, our pursuit of prosperity is rooted in a deep conviction that our brightest days lie ahead. We may not always have an illuminated path to guide our journey. But when we seek the perspective and insight of others, much like I did with Rev. Brian Konkol, we reinforce the wisdom that flows through all living things into actionable foresight.

Planet Pragmatism Provides a Clear Path to Recalibrate Prosperity

The pursuit and definition of prosperity is on the ballot every election cycle. Yet, this past cycle, the stakes feel greater than ever. Americans and all global citizens are undergoing a difficult and demanding process to redefine prosperity in a rapidly changing world. The convergence of advanced technology, including artificial intelligence (AI), with the shifting sands of geopolitical ideology is actively shaping the world, including humanity's vision for "planet prosperity," that is, our intentions for peace, truth, and justice.

Underlying the ever-changing and dynamic political and economic landscape is our relationship with one another and that with the natural world. There are no limitations on how we attain greater prosperity, except those that go against the laws of nature or subjugate our collective humanity and the common good. We must be vigilant in this time of swift change to ensure that all people and voices are seen, heard, and engaged toward where we go from here.

To succeed and thrive, individuals, communities, and organizations must now move beyond the simple monitoring of the temperature (serving as a thermometer) of their key relationships. We must relearn how to listen, learn, engage, and empower — from within their mansion walls and across the most susceptible constituents and challenging stakeholders whom they serve. To succeed in these next few years, we must recalibrate our principles and values with a clear path and playbook for pursuing prosperity with common sense for the common good.

Each chapter of this book will conclude with Points on Pragmatism to provide a summary of the key elements discussed. Below is the first recap for your review.

Points on Pragmatism:

- ***Change can feel intrusive and unwanted.*** *Managing swift and uncertain change requires organizations to relearn how to listen, engage, and empower — from within their mansion walls and across the most susceptible constituents and challenging stakeholders whom they serve. To succeed in these next few years, organizations must recalibrate their mission with their values and their plans for future growth and impact.*
- ***The disorientation of a thick fog settles into all our lives, from time to time.*** *With unwavering faith, codified by our confidence in ourselves and our relationships, we can push through the furor of fog to reveal that the path forward that we seek has always been there.*
- ***The definition and pursuit of a "new prosperity" is on the ballot every election cycle, however, it was felt immensely during the 2024 U.S. election cycle, giving rise to an important inflection point for the future of America, and perhaps the world.*** *Americans and all global citizens are undergoing a difficult and demanding process to redefine prosperity in a rapidly changing world. We must be vigilant in this time of swift change, to ensure that all people and voices are seen, heard, and engaged toward where we go from here.*

INTRODUCTION

Thank you for deciding to peek inside the pages of this book. I'm grateful that you have chosen it, or that it has somehow navigated its way through our incredible Universe and found its way to you. I'm proud to say that this is my fourth published book. My previous titles include the following:

- *The Sustainability Generation: The Politics of Change and Why Personal Accountability is Essential NOW! (2012)*
- *Time to Trust: Mobilizing Humanity for a Sustainable Future (2014)*
- *The Dignity Doctrine: Rational Relations in an Irrational World (2020)*

Whether this book or any of my previous titles are good or useful books, I'll leave it up to you, the reader, to decide. I do, however, sincerely hope that you find a morsel of insight, timely and thoughtful perspective, and new ideas that may provide context for our ever-changing world while also enriching your life and your ongoing journey.

This book, just like my previous three, evolved organically. Friends and colleagues often ask me about the book-writing process. They're usually curious about how I go about conceiving, researching, and writing; how long it takes to complete a manuscript; how to write a book proposal and find a publisher; what else is involved in the editing and publication process; and, of course, whether writing and publishing books is profitable. To this latter interest and inquiry, I'm still trying to answer this for myself. The book business is, as I have come to discover, an incredibly competitive, challenging, confusing, and rewarding enterprise. I have always disclosed, and continue to attest to now, my motivation to write and publish is not for fame, fortune, vanity, or ego. I write to explore, expound, and expand upon existing and new knowledge at the intersection and multi-dimensions of life, work, family, spirituality, and humanity's existence within society and with nature.

As I tell friends and colleagues who ask me about book writing, I have always had an interest in writing, and somewhere in my soul, I thought I might write a book someday, but I never thought my first book would have been published in 2012, when I was in my mid-thirties. I also never imagined that a second book would be published two years later, or a third, six years after that. And now, at the dawn of a fourth book publication, I'm feeling more confident in saying, "Yeah, I write books." There's no substitute for sheer consistency, persistence, perseverance, and grit when it comes to writing — or perhaps many other useful endeavors one can put their time and mind to.

When people ask me where I find the time, my simple response is that I make the time. I know some incredibly gifted and disciplined writers, mentors, and friends. They write daily, as a personal credo, much like practicing a faith, or putting one's energy into exercising, gardening, or anything else that brings joy, even if the gratification is severely delayed. For many, writing is an extension of who they are. The art, craft, and act of writing is also akin to an itch that likes to be scratched but doesn't ever go away. Scratching it can provide momentary relief, that is, a mechanism for self-reflection and expression (aka, getting something off one's mind), but that doesn't always provide the writer with resolve. The itch, for many writers, carries on and continues well after the ideas have been transcribed by words and then further interwoven and refined to create meaning and the power of story. In my case, I write when I have the time, which can be challenging with family, work, a little bit of play, and other extracurriculars that I enjoy, including teaching, mentoring, and working with emerging enterprises. Sometimes the words flow, sometimes they don't. For me, the most challenging aspects of writing and publishing are editing, refining, and simplifying the prose. I've also discovered that I tend to be reserved and conservative in my musings. Perhaps as I grow older, I'll let more of *me* out onto the page. Only scratching that itch over time will tell.

Like my previous three, this book was not conceived or written from a contrived perspective. In fact, following the release of my 2020 title, *The Dignity Doctrine: Rational Relations in an Irrational World,* I took a "long-COVID" hiatus from writing. Honestly, between the Fall of 2019 and Spring of 2024, I did not do much of any writing. I did not feel I could hold a compelling idea in my mind that made sense. Call it the fog of COVID settling in for a long winter's nap, or perhaps I sidelined myself to pursue other important occupations, including work, family, and making sense of what's been happening in the world. With the manuscript for *The Dignity Doctrine* submitted and in development toward publication, I focused on publicizing the book. During this time, COVID-19 rocked the world to its core, and although I had deep thoughts and lots more time, the reality was that I was not feeling the itch, so to speak. The world was overwhelming, and I, like millions if not billions of people, was trying to make sense of it. Although this moment in history provided fuel, flame, and fodder for many writers, I simply did not feel an inner voice that wanted, or needed, to come through. Fast forward a couple of years, and the entrenched politicization of "everything" further reinforced my inwardness. I had and continue to have many opinions on the popular culture and political situation during this time, but as I watched and witnessed the onslaught of talking heads: political pundits, podcast proprietors, current and former global leaders, and "future voices," I felt that the world didn't need another "point-of-view," particularly from me. This said, I know and understand that true leaders speak up and act when they need to. I also know and understand that leaders learn when to be quiet.

The past few years, I felt like it was more of a time to be still, listen, and try to comprehend what the community and world were experiencing in the moment. That moment turned into a few years, and now here I am with a new book, with an itch to scratch and something to say, and delighted that I took my time to properly gather my emotions and thoughts to try and get this book right. Reflecting upon the dark place that I believe most people experienced somewhere in their minds during COVID, I truly hope we never go back there. This said, at this moment, the world is not out of the woods. Geo-political tension and conflict, rampant unsustainable consumption, deteriorating quality of life, homelessness and mental health, unsocial media, broken and dated infrastructure, inequality and inequity, and a host of many other human-induced systemic issues plague society. My intent for this book is not to placate our fears or overstimulate our thinking on the underlying root causes of humanity's problems. Honestly, I think we know them all too well.

There is a bright side if we choose to pause, listen, and learn. Humans are the proprietors of our problems. We can also be the purveyors of our prosperity and saviors of our sustainability. This book is an attempt to get back to basics, to deconstruct the complex mess we've created or find ourselves in, and provide common-sense solutions toward achieving greater prosperity for all. This book was written to offer a perspective on pragmatism, brought forth from the inner wisdom embedded deep within our DNA, that values, including stoicism, restraint, common sense, and the pursuit of truth and meaning, can elevate us to higher standards and quality of life. We don't always have to reach for the Moon. It's there, and it's beautiful. And if you were one of the millions of people who witnessed the 2024 Solar Eclipse, we know that the Moon, Sun, planet, and Universe are also magical. There's so much we take for granted, so much for us to learn and use for the betterment of humanity, and so much more discovery of ourselves and our place in the Universe to be explored and revealed. We just need to give ourselves some patience, time, healing, strength, and rationality.

Sometimes we simply must work with and make the most of what we have. COVID taught us many things about ourselves, our human nature, and our relationship with each other and the environment. COVID taught us that sometimes it is necessary for us to be silent, still, and listen intently to ourselves and the world around us. We need each other, and we need nature. It also taught us that the answers and solutions can be achieved when we are focused, intentional, collaborative, empathetic, and caring.

I'm reminded of an article I wrote and published, one of my last and only pieces published during my multi-year hiatus, right at the onset of COVID. Perhaps the narrative was my self-awareness and self-imposed attempt to hit pause on my writing life, as much as it was an attempt to capture the essence of the heightened emotion of COVID in the moment. I begin the book here, because as the saying goes, "humans have short memories." For a moment in 2020, our human world halted and came to a complete stop. The stillness was eerie and yet electric. The feelings of uncertainty, fear, confusion, doubt, and anger were felt around the world, ubiquitously. These emotions were pushed aside, also momentarily, by compassion, love, and a shared desire to conquer and extinguish the COVID risk.

Nature also seemed still at first, but then it seemed to be recapturing its essence, in our public spaces, which we were occupying much less due to our isolation. For a brief blip of time, the stillness allowed nature to be seen once again. Was nature taking over? Or were we finally patient, still, and reserved to witness and revere nature's beauty, the way we once saw it, innocently as children? COVID was and remains a mass reminder and reset for humanity. Although our memories often repress and push pain points deep into the folds of our brain, we should remember that society and nature experienced something miraculous. For a millisecond of our human existence and planetary time, a singularity occurred where we witnessed the sanctity and interconnectedness of life. We should be more humbled by this.

A Silent Spring Has Fallen Upon Us, It's Time to Listen so That We Can Emerge With Dignity

Originally published, April 15, 2020, Impakter[1]

It has been weeks since I've written anything of consequence, or even released a short story. Since late fall, I've been editing my third book, *The Dignity Doctrine: Rational Relations for an Irrational World,* which will be released this spring, June 9, 2020. The sheer mental energy and constant creativity that goes into publishing a book can be exhausting. With the advent of the coronavirus (COVID-19), however, my creative voice has been quiet and dormant.

Today, my 12-year-old son had a sneezing fit at 2:45 am that woke me from sleep. Concerned, I got out of bed and checked in on him. He was fine. He used the bathroom, blew his nose, and asked me if he could have some water. "Of course," I said. Together, we walked downstairs to the kitchen in the dark. It was foggy outside, a metaphor I thought, for how my mind has felt for the past few weeks. I poured my son a glass of water. He gulped it down with ferocity, taking a deep breath of air afterwards. He seemed refreshed. I asked him if he needed more, he said no.

We walked back upstairs, and I tucked him back into bed. I said, "Back to sleep." He said okay and then added, "Hey, Dad, you know robots?" I thought, oh no, where is this going at 3 am? Reluctantly, I said, "Yeah, what about robots?" My son said, "Well, you know how they feel better after getting some oil to loosen them up? Well, that's how I feel after that water." I smiled and said, "Glad you feel better, now good night." My son returned the gesture, "Goodnight, Dad."

I went back to bed with a smile, and a feeling of warmth covered me. But in the deep silence of the night, thoughts about COVID-19 overwhelmed my mind and became deafening. The more I tried falling back asleep, the louder my thoughts became. I chose to get up, make a tea, and sit alone quietly in my living room. Within minutes, my dormant creative voice had been awakened.

Published in 1962, Rachel Carson's seminal book "Silent Spring" documented the severe ecologic impacts, particularly on birds and their habitats, that resulted from the unfettered use of pesticides in the environment. Carson's "Silent Spring" also unveiled the intrinsic interconnections between the natural and human-built worlds, resulting in a wake-up call for greater responsibility and environmental protections. Carson is widely credited for giving rise to the modern environmental movement.

Today, nearly six decades later, it is estimated that half of the world's 7.5 billion people are in lockdown, sequestered and confined to their homes from the noise and familiarity of daily society as we once knew it. The airplanes are grounded, the transit systems sit empty and idle, and the roadways are no longer congested with the clang and honk of rush hour traffic.

For those of us who have the luxury of working from our homes, the once normalized sound of a daily commute feels as if it is a distant past. In the past three weeks, our lives have been turned upside down, replaced by a surreal reality marked by an unsettling tone and the sobering death statistics announced during daily press briefings. The sounds of spring are often thought of as a time of renewal or rebirth of life. But this year our spring has truly fallen, silent.

It is as if the invisible hand of God has swept its way across the globe, leaving no community or individual untouched by this harsh new existence. This invisible hand is simultaneously reminding all of us of our economic, ecologic, and mortal fragility, as it has, as far as I can see, instilled a newfound sense of humanity and humility within many of us. COVID-19 has been described as a war, with an invisible enemy. The invisible hand of government in the U.S. and worldwide is being forced to become more transparent amid the economic pressures and social needs associated with responding to a national and global crisis.

While the spring has fallen silent, COVID-19 has loudly revealed our vulnerabilities for local and global, urban and rural, resiliency and sustainability. While this was already widely known, it has now become more evident that we've shaped a global economy in a manner that does not optimize resources or equitably value people; rather, it pits human against human, government against government, and humans against nature. This behavior is unnatural and immoral, defying logic and the laws of nature.

Deep down, I believe we've inherently known this, but it has taken a global pandemic to wake us up out of our fog. Essentially, the war we wage is and has always been with ourselves. COVID-19 is impacting everyone on Earth. Although it is alarming and bringing about great sorrow, COVID-19 is also giving rise to a rebirth of humanity.

If we can, for a moment, stop the infighting and quiet ourselves and listen intently to the space within and between the silence, a faint and growing sound of unity can be heard. I hear this sound working from home in between the drips and beeps of coffee makers, tea kettles, dish-and-clothes washers, children's toys, and video calls. A burgeoning sound for change is growing louder each day. It's expressed by local politicians, it's flowing out of chief executives, it's being pleaded for by the brave and selfless doctors, nurses, and other frontline workers who put themselves in harm's way each day.

Humans are determined social beings who seek purpose and direction. The global proliferation of COVID-19 has shaken us to our core. But while we are shaken, our spirit and resolve are not yet broken. Our economy may be on the cusp of collapse, but that can, if we are unified, be redefined and rebuilt, stronger and more sustainable.

Tony Robbins has wisely suggested that humans have six core needs, including the need for certainty and comfort, uncertainty and variety, significance, love and connection, growth, and contribution.

For now, COVID-19 is providing us with a healthy dose of "uncertainty and variety." However, across the other areas of human needs defined by Robbins, there is hope and promise of a new beginning. Together, we can and will reimagine, capture, and create a better and brighter future beyond COVID-19.

With determination and discipline, we will once attain certainty over our lives. We are already taking comfort in our families, friends, and the care that is delivered through the self-sacrificing contributions of frontline workers. Let's use this unprecedented time to do our part by heeding the expert advice of doctors and healthcare professionals. Let's also value our time wisely and use it constructively to redefine what it means to be human and alive on this amazing place of abundance we call Earth.

We can drown the deafening sound of fear and uncertainty with the appreciation and gratitude that we have for each other, through love and connection, as the underlying force that can overcome any barrier or challenge.

I'm reminded by this each day, as I see and hear kids within our neighborhood practice social distancing, but still find unique ways to play, laugh, learn, and create. Imagine, it was only a short time ago when so many parents were pleading with their children to get off devices and go outside for fresh air. Our children are our future.

As parents and caregivers, we need to look beyond our selfish needs during this difficult time and provide our children with undivided love and care, particularly as their lives get redefined. We need to provide them with support, guidance, and quench their thirst to keep their minds sharp, aspirations high, and inspiration flowing. We must also demonstrate strength and be unafraid to show our hearts and vulnerability as we work through this difficult time.

Now is our time to take shelter, but to also take refuge and be open to the silence that connects us all. For it is in this silence that we will not only grow but also discover the common ground by which we rise together to live as one with peace, prosperity, and purpose.

Together, with love and through unity, we are resilient and resourceful. Now is our time to take shelter, but to also take refuge and be open to the silence that connects us all. For it is in this silence that we will not only grow but also discover the common ground by which we rise together to live as one with peace, prosperity, and purpose.

Our spring has fallen. But we will emerge from this crisis with our dignity and a clear direction. It's 5:30 am, I can now hear the birds chirping at the dawn of a new day. Spring has awakened, let's take comfort in knowing that we will soon also rise again.

I've written this book in hopes of continuing my work centered on coaching, mentoring, and developing now and next-generation leaders. A central tenet of my career has been passionately advocating human dignity, sustainable development, and global cooperation. I fundamentally believe that true prosperity stems from planet pragmatism. In the pages that follow, I will provide further definition, context, and insight into that phrase and why I chose it to be the title of this book. This book examines many contemporary and critical topics, including how we can decode humanity's future, questioning if our pursuit of sustainability is doomed, while exploring the moral and ethical dilemmas intertwined with advanced technology such as AI. Let me begin the first part of this book with the underlying greatest opportunity we have for achieving a better, more dignified, and sustainable world — *principled leadership.*

PART I

PLANET

1

HUMANITY BEGAN HOMELESS, WE REMAIN HOMELESS

We were all once homeless; we remain homeless. Beginning from humble beginnings as foragers and then hunters, humans evolved to eventually domesticate the Planet's lands and resources, claiming ownership over anything and everything we saw and could convert into something that enabled our survival and benefited us. If that sounds a wee-bit selfish, well, look around your room at this very moment and then take an honest look in the mirror and ask yourself if there is any item of excess within your sight or grasp. Do you really need everything around you to survive and live a healthy life? Or are some of the objects, possessions, and physical manifestations of our modern world more in the category of "good to have," or wants, as opposed to needs?

I'm not here to judge, as my family and I live comfortably, at least by our standards. If I'm living comfortably, and hopefully you are as well, then why do I begin with the headline, "We Remain Homeless?" Well, after thousands of years of evolution, humans have discovered how to clothe, feed, and shelter ourselves while cultivating communities and our sense of place we call home. But through this process and the years it took to get to the level of comfort we now enjoy, we either inadvertently or intentionally (perhaps a bit of both), distanced ourselves from our one and only (as far as we know) original home, Planet Earth.

Somewhere across the arc of humanity's history, we discovered a concept called ownership. Ownership, defined as the act, state, or right of possessing something, is a complex ideology. The idea of ownership, over ideas, property, and even people, has been at the forefront of human-induced hate, suffering, and destruction for thousands of years. Yet the fundamental construct of ownership is not entirely evil. Like any ideology or institution, the underlying foundation of ethics, values, rules (or lack thereof), and people governing ownership shape how it manifests within society for better or for worse. The privatization of land and its ownership, for example, has had a storied past when it comes to numerous persistent and pervasive social and environmental challenges, including resource management and conservation rights, taxation, social and environmental justice, and more. Today, the rules governing ownership are being used by organizations, including land trusts and conservation groups, to acquire, protect, and conserve wildlife habitat.

The concept of ownership gets to the heart of how the industrialized and developed world operates. Our material-driven consumptive lifestyles are driven by ownership and possession. Our relative status in society is measured by our perceived state of ownership over material and non-material things like homes, clothes and accessories, cars, stocks and financial investments, boats, recreational vehicles, educational degrees, and so much more. Although material possession has been a longstanding cultural norm for much of the developed world, this construct of where society places value on ownership is shifting.

For a diversity of reasons, most rooted in economic pragmatism and some toward environmental pragmatism, new business paradigms have emerged offering citizens and consumers the option to access and live the lifestyle they choose through fractional ownership or even no-ownership. Essentially, consumers have begun to recognize that ownership of material possessions may not be the epitome of achieving a higher social standing or quality of life. Rather, leveraging one's financial ability to access what you want or need when you want or need it can be just as fulfilling, and perhaps without the burden of maintenance and upkeep, depreciation, or unintended risks or consequences.

For example, in 2008, Airbnb emerged onto the scene just as the global financial meltdown and housing crisis occurred. Offering an online marketplace for people to book short- and long-term homestays and experiences, the Airbnb platform provided a new means for property owners to monetize their assets. Today, Airbnb is a $6 billion company that has over 7 million active listings in over 220 countries. In 2022, nearly 394 million nights and experiences were booked on Airbnb around the world. Further, the company has had over 1.5 billion guest arrivals since it started in 2007. Talk about filling a void in the marketplace! Airbnb does not own the properties it advertises and books; the property owners do. This company discovered that there was an enormous marketplace to bridge people who wanted to monetize their assets without selling and those who wanted a differentiated option for short- to long-term homestay experiences (i.e., living the lifestyle, but not owning the asset). Airbnb's story is but one of several that are reshaping the value and values of modern ownership.

Interestingly, another company, Uber, also formed in the 2009 timeframe during the global economic crisis. Rideshare companies, including Lyft and Uber, discovered, much like Airbnb, that an underserved need (and desire) existed between the traditional ownership culture and the emergent "X-as-a-Service" economy (i.e., housing as a service, transportation as a service, etc.) Consumers began to realize that exclusive outright ownership over "X," whether that be a house, a vacation property, a recreational vehicle, jewelry, computers, phones, or even clothing, wasn't necessarily the best value or investment to achieve the standard of living they sought. The advent of the de-ownership business paradigm has enabled millions of people to shift how they put their finances to work in a manner that re-establishes the traditional American Dream from one centered on ownership, to one that is more adaptive and fitting to the way an individual wants to spend their time and financial resources in the here and now.

The De-Ownership Economy: Companies Profiting from Adaptive Consumerism

Company (year founded)	Industry/Service	2023 Revenue ($)	Number of Employees
Airbnb[2] **(2008)**	An American company operating an online marketplace for short- and long-term homestays and experiences. The company acts as a broker and charges a commission from each booking.	$5.99 billion	6,907
Rent the Runway[3] **(2009)**	An e-commerce platform that allows users to rent, subscribe, or buy designer apparel and accessories	$100 million	1,800
Uber[4] **(2009)**	Uber provides ride-hailing services, courier services, food delivery, and freight transport. The company is headquartered in San Francisco, California, and operates in approximately 70 countries and 10,500 cities worldwide. It is the largest ridesharing company worldwide with over 150 million monthly active users and 6 million active drivers and couriers. It facilitates an average of 28 million trips per day and has facilitated 47 billion trips since its inception in 2010	$37.28 billion	30,400

Today, an entirely new economy is being conceived and constructed to provide greater equity, control, and ownership to individuals within society. Later in this book I will highlight the "7 Meta Dimensions," or "7Ds", that ground humanity with principles of planet pragmatism and toward greater prosperity. Three of the "7Ds" include the rise of decentralized, democratized, and digitized technologies and business models. What were merely ideas and frameworks scribbled on restaurant napkins ten or twenty years ago have created a paradigm shift in how we think about the concept of ownership and how everyone can essentially become "an owner." Examples of next-generation technology, business, and ideas that exist at the convergence of a decentralized, democratized, and digitized economy include blockchain technology, Web 4.0, decentralized finance (DeFi) such as Bitcoin, and decentralized energy systems such as virtual power plants (VPPs). The table below provides further definition to these enabling technologies of the new ownership economy.

Enabling Technologies Driving the New Ownership Economy

Technology/ Business Model	Description	Enabling Meta-Dimension(s)
Blockchain	A blockchain[5] is a distributed ledger with growing lists of records (blocks) that are securely linked together via cryptographic hashes. Each block contains a cryptographic hash of the previous block, a timestamp, and transaction data. Since each block contains information about the previous block, they effectively form a chain (compare linked list data structure), with each additional block linking to the ones before it. Consequently, blockchain transactions are resistant to alteration because, once recorded, the data in any given block cannot be changed retroactively without altering all subsequent blocks and obtaining network consensus to accept these changes. This protects blockchains against nefarious activities such as creating assets "out of thin air", double-spending, counterfeiting, fraud, and theft.	*Decentralization* *Democratization* *Digitized*

Technology/ Business Model	Description	Enabling Meta-Dimension(s)
Web 4.0	While hypothetical and undefined by some, Web 4.0[6] represents the next evolution of the Internet, whereby there is greater emphasis on being user-centered, collaborative, and intuitive. Web 4.0 is fueled by and feeds the development of interconnected devices, including through the advancement of artificial intelligence (AI). Some refer to Web 4.0 as the '\Metaverse, although the 3D world is but one evolving element of the Web 4.0 movement. The technological underpinnings and capabilities of Web 4.0 will further catalyze decentralized, democratized, and digitized infrastructure and networks, further supporting ownership economy business models made secure with blockchain technology.	*Decentralization* *Democratization* *Digitized*
Decentralized Finance (DeFi)	Decentralized finance[7] (often stylized as DeFi) provides financial instruments and services through smart contracts on a programmable, permissionless blockchain. This approach reduces the need for intermediaries such as brokerages, exchanges, or banks. DeFi platforms enable users to lend or borrow funds, speculate on asset price movements using derivatives, trade cryptocurrencies, insure against risks, and earn interest in savings-like accounts. The DeFi ecosystem is built on a layered architecture and highly composable building blocks.	*Decentralization* *Democratization* *Digitized*

Technology/ Business Model	Description	Enabling Meta-Dimension(s)
Decentralized Energy Systems, Virtual Power Plants (VPPs)	Decentralized energy systems represent a network of smaller, distributed energy sources such as rooftop solar panels and/or battery energy storage systems that operate close to where electricity is used. A VPP is a system that aggregates and manages decentralized energy assets and serves as a larger "energy plant" that can coordinate with the power grid through software to support behind-the-meter and grid demand and supply. Together, decentralized energy systems and VPPs support the integration of renewable energy generation and electric vehicles. They also enable power grid resilience, reliability, and demand response.	*Decentralization* *Democratization* *Digitized* *Decarbonization*

Some of these examples, such as blockchain technology, provide the foundational architecture and foundation to support new business models, while others, including VPPs, represent the derivation of entirely new ways to design and operate infrastructure that solves challenges and creates value in this rapidly evolving new ownership economy.

This prologue discussing ownership is just the tip of the iceberg on how this human-derived concept has its severe flaws in how it has caused and created inequity, injustice, and conflict throughout the world. We understand that many of the negative externalities associated with ownership are the result of poor policies, governance, and accountability. When the rules of any game favor the rights and privileges of a few people, the opportunity for others to partake and succeed will always be marginalized. However, ownership can, when the rules and governance are grounded by ethics, values, and accountability, result in favorable upside potential and outcomes for a greater diversity of individuals, communities, and new enterprise development.

Most people, thankfully, understand the enormous gift it is to live in a home. Home is, after all, where the heart is. When you think of your home, what comes to mind? For some, perhaps paying for the home and your rent or mortgage is top of mind. For others, perhaps it's those nagging projects you haven't gotten to yet. Sure, homes cost money and require maintenance, but these represent characteristics of having a dwelling that is your own. Homes serve a basic human need and function, which is shelter. Homes in their rawest form protect us from the elements and extreme weather, including wind, rain, snow, sleet, and hail. Anyone who has ever been caught in a torrential downpour without an umbrella, or the punishing Sun on the hottest days of summer, knows the value of shelter. Even the simplest roofs or canopies can be a welcome reprieve during those moments of need. But beyond the basic provision of shelter, homes provide something more.

Back to the question of what comes to mind when you think of home, many people might go beyond the functional and economic elements and elevate the emotional connection they have and receive from a home. Love and joy, peace and serenity, safety and security, a sense of belonging and identity. Homes provide all these emotive characteristics and more. They are like our mothers in that homes provide unconditional love and care. Yes, houses provide a roof overhead, but for most people, they become homes because that is where people choose to seek refuge from the world, nurture relationships, and live their lives. Houses and apartments are dwellings and structures. What makes them a home is the people who occupy them with love, care, compassion, and joy that spills out into every room, nook, and cranny.

Most people choose to take care of their home. They clean, furnish, maintain, invest in, and develop a dwelling that provides for their basic needs and allows them to grow, live, and flourish. Truly, homes are special places cultivated and curated by humans as sanctuaries. Given the deep level of appreciation, respect, and care we provide for our homes — why then don't we provide the same level of consideration towards our one and only, ultimate home, the planet?

Within our homes, we (generally) pick up after ourselves. We take out the trash. We clean and organize. We disinfect and sanitize. We create an environment that serves our needs and enriches our lives. We're busy beavers, squirrels, and birds, burrowing within four walls, storing our food, and making a cozy, warm nest to drift off and dream within every night. Indeed, home can be blissful. But we don't treat our home-of-homes, the planet, the same way. Rather, the planet has been and has largely become our dumping ground.

Our homes are sanctuaries, yet the precious land they occupy is littered with remnants of our consumptive lifestyles. Humans are giving new meaning to the phrase, "sweep it under the rug." Unfortunately, the planet's rugs (air, water, land) have become chock-full of our leftover and left-behind wastes. Plastics, pesticides and chemical compounds, heavy metals, electronics, clothing, biologics and pharmaceuticals, agricultural runoff, and a soup of all kinds of other planet-derived human-made materials that encapsulate thousands of years of human influence. If the planet were our son or daughter's room, we'd say, clean this mess up!

The Planet is roughly 4.5 billion years old. It has had a rocky (no pun intended) existence for much of its history, slowly transitioning from an inhospitable, barren mass circumnavigating the Sun to the incredibly diverse ecosystem of life that it is today. Human influence and impact on the plant have been significant. Like a young child in their room or backyard exploring the world around them, humanity has been hard at play on the planet. We've thrown our toys around the room, tracked soil and debris everywhere, and spilled our juice on the carpet. The problem is, we haven't had a parent guiding us or telling us when and how to behave. Yep, humanity has played hard in an unkept room. Our smelly socks have been strewn across the scenic landscape. But there is hope for the planet and for humanity.

My sense after living for nearly half a century on this planet is that humanity is continuing to discover, learn, grow, and mature. For some, this belief may be a bit altruistic and optimistic. In recent years, demagogue politics and the heightened politicization and polarization of everything have left many people fearful, confused, uncertain, angry, alone, marginalized, and a host of other not-so-good feelings breeding skepticism and disdain. This is sad and unfortunate. But, as dreadful as the present state of the planet may seem, the optimist and pragmatist in me believe that what we are experiencing are the growing pains of a society growing from toddler to adolescent and young adult.

We are not out of the woods yet, and we continue to trudge through the muck of our past as we learn to live more responsibly and accountably in the present. We have become more astute to our influence on the planet. We've recognized that Earth is home and that it should be valued and cared for just as we are for our domestic homes, with love and gratitude. We're still learning to listen and understand, to each other and to nature and the Universe. We're only just learning to pick up after ourselves. Growing up is hard to do, and we have a long way to go, but it's time and it's necessary to finally accept and love this planet as home.

Points on Pragmatism

- *Planet Earth is our home. We should value and care for the Planet just as we do for our domestic homes, with love and gratitude.*
- *Humans have a complex relationship with ownership. The concept of ownership, over ideas, property, and even people, has been at the forefront of human-induced hate, suffering, and destruction for thousands of years. Yet ownership has also been a tool that has supported environmental protection and conservation. There are new and emerging models of ownership that seek to create greater prosperity through the opportunity of shared ownership.*
- *We're still learning to listen and understand, to each other and to nature and the Universe.*

2

WHY WE NEED TO ENSURE THAT PRINCIPLED LEADERS AND LEADERSHIP PREVAIL

Why We're Missing the Point (and Opportunity) on Leadership

As I write this book, the United States is smack in the middle of a presidential election cycle. That's not new news, and I promise that I am not going to pontificate about picking favorites. The rising temperature and unease of this presidential election cycle are, however, illustrative of the unrelenting and unforgiving times business, government, and civic leaders are now navigating through.

Fear and frustration over rising inflation, job security, and quality of life measures, including accessible and affordable healthcare, housing, education, energy, transportation, and food, have millions of Americans yearning for new leadership. Further, as partisan politics has ratcheted up a culture of divisiveness, discontent, and distrust among many Americans, we've found ourselves in a lose-lose battle of who's right and who's wrong. Meanwhile, our society remains starved for authentic and principled leadership across all levels of major institutions, including government, corporate, media, academic, research, medical, and non-governmental organizations.

Look no further than the ongoing Boeing and airline industry safety debacle; or the slate of dizzying daily headlines that reveal the lack of leadership that plagues our world. Wildfires, shootings and assassination attempts, escalating global conflicts, contaminated water and foods, homelessness and housing crises, mental health crisis, the barrage of Breaking (and bad) News goes on and on.

I'll stop there. Reciting our challenges with no solution is futile and a waste of your and my time. We all know, or at least have a strong sense, that something is dreadfully wrong; our systems are failing us. We must also consider that we are failing our systems and, subsequently, ourselves. We're created and are now caught up in our own paradox of limitation, constructed upon our false notions of who leaders are and what good leadership looks like.

Great leaders and leadership do not operate in a vacuum. This is self-evident, however, it seems that many people associate leaders as if they are mythical beasts, like Charizard, one of my son's Pokémon playing cards. The beast has unique powers and can miraculously manifest those powers whenever it needs to. American culture likes to put leaders on a pedestal, celebrating those who fight, those who win, and those who personify strength and might. Deep down, most people understand that leaders are not mythical beings. Society's caricature of leaders and leadership is an overglorification of simplistic traits that paint a picture of masculinity, dominance, and power.

Well, I'm sorry to burst this bubble, but those traits are not what most leaders are, and certainly not what leadership is all about. Further, the challenges and opportunities we have before us across all facets of society require a much different type of leader, one that challenges our conventional views on leadership, and one that is much more caring, accessible, inclusive, humanistic, accountable, principled, and trustworthy.

Our shaped cultural perception and misconceptions of leadership have misguided many Americans on what leaders should look and behave like. Some people have accepted the brute and simplistic noise makers as images of power. But you cannot trust those who pontificate the desire for power as a "trust me" platform. Trust is a relationship; it requires all parties to put in the necessary time and effort to pursue and achieve something together. Trust is not one-directional, therefore, leadership based on trust must be multidimensional.

Leaders must be supported. When things are not working at your job, in your community, within your home, it's easy and convenient to point your finger at someone or something else as the problem. High inflation? Chastise the President. High Crime? Oust the Mayor and Chief of Police. Frustrated about your job performance? It must be because your manager doesn't "get you" or coach you enough.

Inherently, most people have been conditioned to place blame and frustration on someone or something with higher title and authority when things aren't going their way. Many people don't understand or at least don't want to put in the time and effort to understand that leaders and leadership are not (exclusively) about titles. Rather, leadership is about creating and nurturing a trust bond where all people have a role to serve in the maintenance and longevity of that bond. If all parties are not actively engaged, something always falls by the wayside and is lost. For some, it is far easier to attach themselves to conspiracies that serve their narrative than it is to serve their role in manifesting trust.

Our economy and society may or may not be "rigged." What is certain is that if we don't solve the leadership dilemma, we are destined to find ourselves in an increasingly polarized, unproductive, and manipulated economy and future. To be clear, I'm not suggesting that all people carrying a title of significance are poor leaders. I'm also not claiming that leaders don't exist outside of formal titles, like President and CEO. They do.

Leaders exist and leadership is exercised across all organizations and institutions, and in a myriad of ways. In recent years, anyone with a leadership title has become a more visible target, particularly for those who epitomize leaders as the mythical creatures lurking atop the pedestal. We've witnessed an increased assault on people who hold title, particularly as more and more people have become empowered to speak out against anyone and everyone who does not align with their point of view, their politics, or their values.

Exercising First Amendment Rights has always been an American virtue. But the demeanor by which some individuals have chosen to exercise those rights in recent years has become noisier, pugnacious, and outright disgusting. If we view leadership as a trust bond, then when it comes down to it, we are all culpable for poor leadership. It may be true that some "leaders" warrant harsh criticism. And some "leaders" simply do not have the skillset or experience to build trust and lead. Leaders must be willing to self-identify their weaknesses or gaps, or at least be willing to listen and hear what those limitations may be from those they serve, and then learn and grow, adapt and evolve, or accept and move on. With all the noisy oscillations in the world that seek our time and attention, it has become more challenging for us to listen with intention and toward understanding. The *sidebar* below examines this phenomenon further.

Leadership is not about placing all responsibility and accountability with one singular leader. To be effective, leadership must engage, empower, and envelop commitments from all whom it serves. In this way, we all have a role to serve as active participants in the institutions that we place value with, such as family, religion, democracy, capitalism, education, and more. No singular institution or singular leader is responsible for all our prosperity. Prosperity, like democracy, is valued and earned through our active participation, that is, by being a leader among many leaders.

From Nonsense to Clarity: Professional Growth and Leadership Development Require Us to Sharpen Our Focus and Filter Out the Unproductive Noisy Oscillations of the World

Our digitally interconnected world has become chock-full of manufactured noise. As I write this passage, I hear a clock ticking beside me, the wind rustling through the newly sprouted spring leaves, the distant hum of the road as cars and trucks drive by my home, birds chirping as they celebrate the rebirth of nature, the consistent tumble of the clothes dryer, and the clickity-clack of my laptop keystrokes. Most of these sounds are mostly pleasant, if not hypnotic. Some of the noises are natural. Others are completely manufactured from human products and systems.

I have the television turned on, but it is muted, as is my phone. It's a relatively peaceful and quiet day, but the noise in my head is nagging and persistent. The noise cannot be audibly heard, yet it's there. To be clear, I'm not hearing voices or other sounds within my head. I am filtering and buffering a buildup of external nonsense that has formulated a frequency or sine wave of bundled thoughts bouncing around my brain. I'm attributing my inner thoughts as sound here because they represent and are producing, at least in my mind, unfiltered and excessive noise. If not monitored daily, thoughts can increase their frequencies and become noise. Our thoughts stem from all kinds of stimuli. Our hyperconnected digital society has amplified the sheer number of external media and inputs that penetrate our conscious and even subconscious mind.

There is good noise and bad noise. The rhythmic sounds of nature, music, or even a loved one's voice, all may be soothing. But nature, music, and the reverb of that loved one's voice can also produce deafening noise to the brain under certain conditions. Some people find comfort in the industrial sounds of a large city. The chatter of people, thud of construction, wail of the siren, clank of the subway, hum of the building systems, and honk of the horns can feel as if they are being conducted, much like a symphony. Separate and single out any of these sounds, and they are quite annoying. Put them together, however, and there is a beat and a harmony that comes through. It's strange and yet beautiful at the same time.

I'm referring to "noise" not only as an audible and heard input, but also as external media and inputs that engage our senses, including what we see, taste, feel, and experience. While we may assign visual and audible elements to our digital society (i.e., watching screens), we are equally engaging our other senses as we do so. The "noise" or chatter that builds in our minds is not just a remnant of what we've heard, it is also heavily influenced by what we see, smell, feel, and interpret through our intuition.

Noise can be soft and low-pitched; it can be loud and high-pitched. Noise can be soothing; it can also be chaotic and confusing. Sight, sound, smell, and our other senses shape our thoughts and experiences; they also provide a medium for our interpretation of the world around and within us.

Speaking of noise, our 24-7 data-infused social-media-crazed society certainly has a lot to say. Or does it? Sometimes I catch myself scrolling social media, during the day, in the evening, in the morning, and pull myself by the back of my neck out of the slippery slope into the abyss, only to ask, what was I doing? What was I searching for? What was I watching, reading, listening to? Was I conscious? Am I passively or actively engaged in content? And to what end? We are, if we choose to be, constantly bombarded by sounds and images curated by others and the rhetoric delivered by the media, politicians, and other personalities.

Our external world, that is the world that we allow to infiltrate our minds and consume our thoughts, is like taking a vitamin. A daily and measured dose is healthy for the body. However, if we infuse our minds and bodies with too many vitamins, they can build up in our system and be toxic to the liver, kidneys, and other critical organs. Given the advance of "everything digital," perhaps a daily dose of external media and input is necessary for us to serve as functioning members of society. Whereas people used to get a lot more information from newspapers and magazines, today, information flows more instantaneously into the palm of our hands. This said, not all information is good information.

The advent of AI has promulgated ethical and moral issues, including the propagation of misinformation and the use of deepfakes, among other privacy and personal data protection concerns.

Information today is conveyed much more visually than ever before. I remember, as a child, looking at pictures in the newspaper, reading the captions, and feeling connected to the story. With instantaneous communications and videos capturing life in real-time, today, we are part of the story. It's a different world, and we must acknowledge that we have a choice on when, where, how, and why we choose to engage.

Much like the vitamin metaphor, tempering our intake of daily media may be part of a healthy regimen. When we overconsume, our heads can quickly be filled with noise. Although our minds are powerful, they are not meant to take on the mass inputs that we currently do through our digitally connected world. Our mental acuity and personal time were not meant to be fixated exclusively on data and screens, studying the beautifully curated lives and existence of others, or absorbing the pain bodies that lie within the ugliness of society. Thus, it pays to be aware of what "noises" you are choosing to let infiltrate your mind, body, and spirit — and assess what impact they have on your thoughts, your health, and your overall well-being.

When we limit unnecessary noise in our lives, we choose to be more present with what shapes us, our thoughts and intentions, and ultimately, who we are. Take a moment to monitor the noise around you. Go beyond what you hear and examine what you see, smell, feel, and if possible, what your intuition and gut are telling you. Take a few deep breaths. Try your best to calm your breathing, relax, and sit with yourself for a moment of conscious self-awareness and observation. Take in, through your senses, all the good and bad noise.

Can you filter the noise? Are you mindful of your thoughts, breathing, and state of being? Practicing this as a daily routine can help build and condition your noise filtering function so that you can eliminate "nonsense noise" and feed your mind, body, and spirit with the appropriate dose of noise that provides clarity and focus to your life.

Try to maintain a consistent routine for noise awareness and filtering, and test out different techniques that work best for you. Recently, I stopped listening to the radio in my car when driving. I've found that by practicing noise filtering, I am more attuned to the pleasantries of the good noise that life has to offer. When we layer noise with noise, our mind's capacity to process (let alone live within) the moment becomes limited. Filtering excessive inputs of noise enables us to center our mind and body in the now, enriching our conscious experience and ensuring that what we're recording within our unconscious mind leads to a healthier state of self, including more productive thoughts, actions, and behaviors.

Now get out there and make some ***good*** noise!

A New Definition of Prosperity is Emerging, Requiring a New Kind of Social Leader in All Sectors and Institutions

Once a beacon of prestige and freedom, the classic metaphor for pursuing and attaining the American dream has eroded. Decades in the making, the demise of the American dream represents a confluence of economic, environmental, and social forces that have been eating away at people's quality of life and their pursuit of prosperity. Although the romanticized American dream has faded, a new generation of Americans and global citizens are seeking to define their fate and freedom toward a new prosperity. Fed up with big business, big brother, and big tech, citizens and consumers feel as if they are trapped living within a world of deception, delusion, and distrust — some might even associate the dissolution of the American dream to the advance of the dystopian state captured within George Orwell's *1984*.

The American society, and democracy at-large, has long had to fight for survival. We've been shaken by a steady state of social change and civil unrest that some historians and pundits postulate as the growing pains America and Americans have had to go through — as democracy and freedom is never simply afforded, they must be intentionally pursued and protected.

The origins of a new prosperity are underway. Guided by redefining prosperity (the underlying "why") and through principles of pragmatism (the "what and how" of the new prosperity playbook), people have shed notions of enslaving themselves to institutions and systems that permeate false beliefs and narratives around freedom and prosperity. People are reclaiming their virtue and assuming control of their future. While a rebellion of ideology may ensue, if not a more pronounced rebellion that could arise, my sense is that individuals and our broader society are yearning for a greater sense of certainty, belonging, identity, hope, and love. This is not to say that we will not continue to see social uprisings or heated debate in the coming months and years. Given the climate we're living in, we are literally and figuratively, as Thomas Friedman's book title connotated, living in a *Hot, Flat, and Crowded* world.

Truly, the divisive and heated political temperament seems to rise like that of the planet's temperature. Our resources are limited, yet we continue to be wasteful. Although the corporate sustainability movement is visible and billions of dollars flow toward "sustainable growth," billions of people around the world live in dire conditions, fighting for their survival. The movement toward a new prosperity has mobilized. In the next decade this movement will rethink and redefine our notion of wealth, ownership, status, and power. In doing so, this movement will become a new hope toward a more peaceful, just, equitable, dignified, and sustainable society.

Great Leaders Understand that they are Entrusted to Serve

Great leaders are human, entrusted to serve at the will of their constituents (employees, shareholders and investors, citizens and voters, trustees, family members, etc.). President, CEO, Chairman, Executive Director, Principal, Superintendent, Head of XYZ, whatever the title, these individuals were elected or selected to be the bearers of a mutual trust. When those they serve no longer entrust, the leader's power is relegated to a functional title. Unfortunately, this is when many ego-based leaders double down on poor judgment and behavior. When their leadership is challenged, they dismiss the trust bond and sanctity of the relationships of which they were chosen to nurture, resulting in their continued demise. The underlying erosion of trust leads to a breakdown of relationships that upheld the institution, at least until a new leader can emerge to provide the necessary trust bond.

People only have as much power as they are allowed to have, bestowed by the trust bond provided by others. Leaders recognize and respect that the [perceived] power they have is not to be wielded with arrogance or greed. Leaders understand that their influence and impact are gifted by those who choose to be a part of something greater, and founded on evergreen principles of honesty, integrity, trust, accountability, humility, dignity, and more. When it comes to our leaders, business, political, religious, civic, and so on, let's stop setting them up for failure, and us for disappointment and frustration.

As we pursue a refined vision for what American prosperity should be, let's take a hard look at the type of leaders we need and what leadership should be all about. We must shed our past impression of leaders shaped by mythical qualities and cultural icons. We must also embrace and lean into the fact that leadership is everyone's responsibility. For America's business and political leadership to grow and lead, we must all pitch in. Leadership is about providing an opportunity for all to learn, grow, and lead while being active trust brokers and bearers in achieving a more prosperous future together.

Points on Pragmatism

- *We're Missing the Point (and Opportunity) on Leadership*
- *A New Definition of Prosperity is Emerging, Requiring A New Kind of Social Leader in All Sectors and Institutions*
- *Great Leaders Understand that they are Entrusted to Serve*

PART II

PEOPLE

3

SUSTAINABILITY IS A HUMAN CONSTRUCT AND PLANETARY PARADOX

To begin with, let me say that I am a staunch advocate for sustainability, sustainable development, sustainable education, enterprise, energy, food systems, infrastructure, really anything and everything, sustainable. The foundation for sustainability, including requirements for long-term, intergenerational, holistic, and systems-level critical thinking, has always intrigued and inspired me to "dig in" on everything sustainable. The transdisciplinary nature of sustainability has always been intuitive for me, placating my desire to integrate wisdom, knowledge, and experience into innovation and solutions for solving the problems of today, and alleviating the problems that may arise tomorrow.

For more than twenty-five years, I've been on a journey to do just this, working to solve "the problems worth solving" by integrating multidisciplinary expertise through a lens always focused on understanding and fostering sustainability, in a practical and sensible way. I've worked as a sustainability researcher, educator, writer, speaker, business advisor, and practitioner. As a practitioner, I've had the privilege to work within and across a diversity of sectors and enterprises, including applied research, government, early-stage entrepreneurial, manufacturing, engineering, and other commercial and industrial organizations. This experience has been incredibly insightful and rewarding. Many times during my career, I've reflected upon the fact that I get paid to learn, grow, adapt, and innovate. It has been a wonderful journey, and I continue to chart new courses of action each day.

With this disclaimer in mind, let me offer a perspective that I believe is informed by a dynamic career that has, like others, seen the highs and lows of personal growth and development, the fun and frustration of different types of organizations and work, but all-in-all a great deal of relationship building, learning, discovery, and professional growth. Some observations and lessons include the following:

- Sustainability is dynamic, always shifting with nature, and with the nature of society; Currently, we are not sustainable.
- Sustainability is more than the sum of its parts.
- Sustainability has become overly politicized.
- Sustainability can be subjective, so *Start with Why*.
- Sustainability is the value, not the solution.
- Do less harm and do better together.
- Sustainability is akin to the human body's critical systems.

I will now attempt to provide some thoughtful context and insight into each of these observations and insights.

Sustainability is dynamic, always shifting with nature, and with the nature of society; Currently, we are not sustainable.

A classic definition of sustainable development, grounded by the 1987 United Nations Brundtland Commission[8] defined it as "meeting the needs of the present without compromising the ability of future generations to meet their own needs." Note that this definition does not say anything about specific goals or targets, such as greenhouse gas (GHG) reductions, specific waste reduction targets, or whether you should use paper or plastic. The Brundtland Commission definition does provide three distinct characteristics of sustainable development. First, it references the present generation's needs. Secondly, it references the needs of future generations. Third, inherent in exploring the present and future is the requirement of time.

Given this longstanding definition for sustainability, right now, here, and today, I feel confident in writing that we are not sustainable. I believe that while it is well-intentioned, the Brundtland Commission definition for sustainable development is misleading, or at least it requires us to calibrate the definition according to our existing situation. Let me explain. I would argue that most humans on the planet are not having their basic needs met. Although extreme global poverty is declining, currently 9 percent of the world[9], or nearly 712 million people, live below $2.15 USD per day, characterized as extreme poverty. Hundreds of millions more people also live vulnerably close to this threshold, also struggling to nourish their bellies, let alone have the proper access to basic healthcare, education, energy, and infrastructure. If we are unable to meet our basic needs today, how can we possibly be focused on the needs of our future generations? I believe that many sustainability "leaders" and practitioners have long been under a spell, me included.

The allure of the Brundtland Commission definition, for many "first world" sustainability practitioners, is in the future tense, "without compromising the ability of future generations to meet their own needs." For those who have met their basic needs, it can be far easier to focus on planning for the future. The "first world" sustainability practitioner movement has succumbed to the allure of corporate sustainability and elevating sustainable enterprise, including entrepreneurship, as a pathway for achieving sustainable development. I have worked and lived in this space for twenty-five years and love it. Looking back at my time and career, however, I cannot help but acknowledge that I have been biased in my romanticized assessment of how to embrace sustainability in life and career.

There is no doubt that there are plenty of sustainability challenges and opportunities and work to do within developed countries, and across corporate, government, research, defense, and other first-world organizations. It can feel shallow in realizing that tweaking out another unit of efficiency for a food manufacturer to make more profit as they seek to reduce carbon emissions does not necessarily impact billions of hungry souls. Stretching the food supply with nominal efficiency gains will not get us to sustainability, particularly if we spend more time, energy, resources, and intellect on the present tense of the Brundtland definition, "meeting the needs of the present." Most know we live in a world of inefficient distribution of wealth, resources, goods, and services. And most understand that geopolitical, cultural, religious, and militaristic forces also influence resource production, consumption, and utilization. From a planetary systems perspective, we cannot focus on only the present or the future state of humanity's existence. We must do and execute both well. This means that we will have to challenge the paradigms of growth, development, political and regional power, institutional hierarchy, market forces and barriers, individual and human rights, and a host of other considerations if we really intend to immerse people and planet in a real conversation around sustainable development.

Sustainability is fluid, in that each generation has a right and an obligation to evaluate, take stock and assess, and determine what its most pressing risks and challenges are, and to address those in a responsible and accountable way. Hopefully, our predecessors provided us with a foundation that can be improved upon. And hopefully, they provided us with enough resources and transferable wisdom and intellect to advance the plant and society forward. But we also know that our planet, our home, is not static; that is, the planet is continuously changing. Sustainability then also needs to envelop the elements of time, continuous change, and a preparedness for the unknown.

There is also the popularized seventh generation[10] connotation of sustainability, which states, "In every deliberation, we must consider the impact on the seventh generation." And then there is the

United Nations Global Compact and Sustainable Development Goals (SDGs), which provides frameworks for business, government, and society to act on more specific sustainability targets. The UN states[11], for example, that "the Sustainable Development Goals are the blueprint to achieve a better and more sustainable future for all. They address the global challenges we face, including those related to poverty, inequality, climate change, environmental degradation, peace, and justice."

The UN Global Compact[12] calls itself "the world's largest corporate sustainability initiative." Grounded by ten principles, on human rights, labor, environment, and anti-corruption, the Global Compact is an initiative that works with companies that voluntarily choose to align their corporate strategy and operations with pragmatic actions that serve to advance societal goals. Through outreach, education, peer-to-peer learning, and engagement, the UN Global Compact seeks to aid companies in understanding and distilling the UN's 17 SDGs into more concrete business priorities, goals, and objectives. In this way, the UN serves as a convener and connector between the industrial community and the diverse social, economic, and environmental needs of the planet and society. This background on sustainability is not entirely complete. Further study and investigation into the culture and history of indigenous peoples of the world and human civilizations; the rise and fall of societies, including past and present economic and governance structures; and, the introduction of modern sustainability frameworks (i.e., Triple Bottom Line, Natural Capital) informed by the individual and combined works of a host of philosophers, writers, economists, mathematicians, physicists and so many others; each contain significant knowledge, insight, and contribution to the current state of thinking and evolving ideology and practice of sustainability.

My intention is not to provide a historical summary or bibliography of this enormous foundation. One could argue that the current state of society is an agglomeration of ancient wisdom and insight. Our current society is made up of a mosaic of wisdom, governance structures, spiritual contexts, cultural identities, and norms. Much like how the diversity of the planet's ecology contributes to the unique richness of life, the diversity of humans and our existence is interwoven with the planet's history and the evolution of its ecology. Our interconnection with each other and with nature illuminates another level of embodied wisdom that, if celebrated and harnessed with dignity and respect, can yield a truly symbiotic and beautiful existence.

Sustainability, then, is a construct that requires input, consideration, participation, and engagement from all peoples and all living things. People are not separate from nature. Nature and people are not separate from the planet. And people, nature, and the planet are not separate from the vast space and Universe in which we occupy. Thus, we are all interconnected across space, time, and certainly in the moment of existence that we share. We live in the same proverbial house. We share the same ceiling, walls, doors, and walkways. We breathe the same air, drink the same water, and feast from the same bounty. Our home is big. In some places this planet can feel isolated and lonely, yet in other places it can feel crowded and overwhelmed. Humans place limitations on our home, on nature, and on our capacity to thrive collectively, and on each other, individually. These limitations are false and grounded in our cultural beliefs, psychology, sociology, and egoism. We have the knowledge, tools, and wisdom to be better, do better, and achieve better, in the here and now, and for future generations.

Given the state of geopolitical affairs and the construct by which society currently functions, there is no singular framework, policy, technology, or act that can "solve for X," that is, fully address all stakeholder interests while providing a solution to a specific sustainability challenge. One could argue that zero consumption could be a pathway, but this is not a realistic argument given how the needs of society are currently served by institutional and economic forces. The underlying values and belief systems that are entrenched within global institutions, peoples and cultures, and ideologies of government and commerce do not provide the same level of urgency or care to sustainability concerns. Accordingly, human networks are continuously pushing and pulling for resources, relevance, and responsibility. Achieving shared responsibility, although noble and prudent in the context of global sustainability, remains as elusive as spotting an endangered Sumatran Tiger in the wild.

Sustainability is more than the sum of its parts.

In recent years, sustainability has become synonymous with topics including decarbonization, electrification, ESG (environmental, social, governance), "net-zero," nature-based solutions, science-based targets, carbon neutrality, and a plethora of additional technical and popularized terms and language that has been used to draw attention and action toward the planet and society's environmental, social, economic, energy, and governance challenges. The mainstreaming of sustainability as a catch-all term, used ubiquitously by academics, engineers, politicians, citizens, and consumers, is fraught with challenges.

Sustainability has become overly politicized.

In fact, sustainability has become polarized and politicized largely because many practitioners have used the word lazily, as a proxy for other words and phrases that can be more descriptive, purposeful, and grounded in addressing challenges and creating value. Sustainability practitioners have used jargon that has compounded the confusion and complexity of how sustainability is thought about, often resulting in polarizing points of view or indecision. This has created a real but also artificial and unnecessary tension between stakeholders. More stakeholder engagement, interaction, and shared action are required to shift society toward more sustainable behaviors. But having greater stakeholder engagement for engagement's sake is not enough. We need to work to understand the nuance of language, the diversity of people, places, cultures — and work toward a common understanding of just what it is we're talking about when we talk sustainability.

Sustainability can be subjective, so Start with Why.

There is a difference between humanity's relationship with each other, the planet, and the multiverse of pathways that can be pursued toward sustainability. For the past six years, I've had the pleasure of teaching Sustainable Enterprise to undergraduate and graduate students at Syracuse University's Whitman School of Management. Each semester, I begin with an exercise inspired by author and inspirational speaker, Simon Sinek's book *Start with Why. How Great Leaders Inspire Everyone to Take Action (2009)*[13] In this exercise, I simply ask students to review and explore what Sinek calls the "Golden Circle" of value creation. Sinek's circle is comprised of three rings where the outer ring is characterized as *What*, the middle ring as *How*, and the inner ring as *Why*. Sinek postulates that enterprise, and its employees, typically know, and can articulate with detail and precision, *what* they do. Sinek then claims that a fewer number of employees can fully describe *how* they do it (i.e., describing the business model, the functional elements of how value is created and delivered to the customer). And finally, Sinek astutely observes that very few can clearly articulate *why* they exist and do what they do to create value.

Sinek equates the Golden Circle to the human brain's neocortex and limbic system where the neocortex regulates our ability to have rational thought, analytical thinking, and language (hence *What* we do), and the limbic brain support our gut instincts, that is, our ability to trust, think critically, evaluate logic and reason toward decision making purposes (hence, *How* and *Why* we do things)[14]. At the center of Sinek's Golden Circle is "Why," which interestingly correlates with the limbic system and our ability to express our motivations, beliefs, and behaviors. Sinek believes that great leaders and organizations communicate from the inside out. They communicate their values, beliefs, and motivations first, and then get to How they do it (process) and What they do (to accomplish a result or impact).

A few years ago, I adopted and adapted Sinke's Golden Circle to my teaching of Sustainable Enterprise so that students can investigate how powerful value propositions (Whys) can, with disciplined and focused business models (Hows), can achieve sustainable impacts (Whats). I provide examples of business leaders and enterprises that inherently communicate from the inside-out effectively, such as Yvon Chouinard[15] of Patagonia, Hamdi Ulukaya[16] of Chobani, Sara Blakely[17] of Spanx, and Jessica Alba[18] of The Honest Company. This exercise is effective in helping students understand the power of communicating a compelling "Why," and particularly for enterprises that seek to create social, environmental, and economic value for their customers. I also leverage this exercise and Sinek's Golden Circle from an individual development perspective. For undergraduate students, I ask them to write about their "Why" at the beginning of the semester, trying to capture what motivates them and their values and sense of value. Then, at the end of the semester, after weeks of immersive learning and engagement about sustainability, I ask students to revisit their "Why" in a follow-up assignment where they can choose their medium of expression (i.e., essay form, video, presentation, or others). Some of the students have gone on to publish their perspective on Ecology Prime, a global organization I'm affiliated with whose mission[19] is to "advance the study and embrace the exploration of the environment and environmental dynamics through global engagement & communications, collaborative education, eco-exploration, travel and publishing for every person … locally, regionally, worldwide."

It is fulfilling, on a selfish and personal level, to see the progression that occurs related to how students characterize their "Why" over the course of fifteen weeks. From all that I have seen from students "Why" assignments over the years, I remain highly optimistic about the next generation's sense of their fate and future on the planet, and their personal motivation and intention to lead a meaningful and impactful career and life. In 2012, when I wrote and published my first book, *The Sustainability Generation,* little did I know how relevant and evergreen that title would remain many years later. I continue to see a 'Sustainability Generation' rise within and uplift society, choosing to shape our future, proactively, productively, and pragmatically — driven by our spirited and smart youth who have an enormously powerful sense of their Why.

Each semester, I also conduct an exercise with undergraduate and graduate students that I've termed "Remove yourself from the 'S' word." During this exercise, I ask students to break into teams to discuss sustainability, and then to report out on what they believe are the underlying (shared) values, elements, and examples of, and more clearly defined descriptors for, sustainability and sustainable enterprise. I ask the students to engage with their peers and to provide their report-out to the broader class without using the "S" word, sustainability. The reason I continue to introduce and conduct this exercise with students is that the S-word has become commandeered by popular media, business, government, and society to mean different things. As a result, sustainability has become both popularized and politicized, a paradox that challenges practitioners from working toward rational solutions to challenges we currently face. I often tell students that even I, as an educator and practitioner, have been part of the problem and have, at moments of either laziness or indifference, spoken about sustainability as a ubiquitous and all-encompassing ideology, practice, business strategy, and societal goal.

Sustainability is the value, not the solution.

While I believe that conditioning society on the difference between unsustainable consumption and sustainable development is necessary, I've always also believed that sustainability should not be mandated, dictated, or overly regulated to accomplish a sustainable outcome in the current moment, or to secure a sustainable future for humanity. Each generation will have its own set of values and beliefs, drivers and trends, goals and priorities, capabilities, and resources. As such, sustainability, broadly, should envelop a long-term generational point of view, but also be grounded in the pragmatism of what's possible (to be evaluated, designed, enacted, or accomplished) within any current generational context. In recent years, sustainability has been entangled with a host of "sustainable solutions" and pathways (often which are policy-oriented and overly politicized) which can be confusing, misleading, and outright illegitimate. For example, phrases like sustainable energy, sustainable transportation, sustainable cities, sustainable infrastructure, sustainable business, sustainable aviation fuels, sustainable buildings, and so on, are not overly helpful in that they are broad in their orientation and subjective regarding their interpretation, application, and desired impact across a short, mid, and long-term time horizon.

To further illustrate the distinction that's required when one discusses sustainability, I point to the need to better define and describe the goals and objectives, utility and function, timeframe (short, mid, long-term), and intended known versus unknown impacts and possible risks or consequences of any "sustainable" action. The idea of "human engineered solutions" as a construct versus sustainable "X" (where X could be energy, mobility, buildings, fuels, homes, technology, etc.) begins to provide more clarity on the value creation that can be designed, implemented, and captured in the here and now, pragmatically, versus what the longer-term impact might yield. For example, renewable and clean energy technologies and solutions, including solar and wind, battery energy storage systems, microgrids, geothermal, hydrogen fuels, and fuel cells, are often characterized as sustainable energy solutions. These technologies offer significant opportunities toward supporting specific and important attributes of sustainability, such as greenhouse gas (GHG) reductions, lower carbon energy, and distributed and resilient sources of energy. But each of these technologies and solutions, even though they may be displacing dirty fossil-based fuels, also contributes to social, ecologic, and economic externalities, which negatively impact the environment and public health. These include the sourcing of rare earth minerals (conflict minerals); utilization of fossil-based energy to stimulate their manufacture, distribution, and installation; and operating, maintenance, and decommissioning requirements that also require inputs of energy, chemicals, water, and other resources that contribute to having an impact on the planet and its people.

Do less harm and do better together.

At this point in our shared existence, humans have generally touched every part of the planet in some way, shape or form. Humans have been consumers, and in the past two centuries we have dramatically compounded our appetite for consumption. When I think of sustainable products and technologies, and broadly, "solutions," I often think of the phrase, "do less harm and seek too always do better, together." I'm convinced that there is true "net-zero" impact solution to solve for "X" that exists. The ultimate answer to that question would require society to have complete abstinence from any type of development, economic growth, or consumptive-based activity. Although some would deem this to be ideal, the reality is that a zero-growth society is impractical and unreasonable — notwithstanding some miraculous technological invention. We know that there is no "silver bullet" sustainability solution, that is, one singular technology or pathway that achieves the paradox of continued economic growth with zero negative impact on the environment, human health, or society. Achieving economic prosperity with as minimal an ecological footprint as possible, and in a manner that improves our quality of life equitably, is proving to be a challenging goal for our current society and institutions.

Sustainability is akin to the human body's critical systems.

Integumentary System, Skeletal System, Muscular System, Nervous System, Endocrine System, Cardiovascular System, Lymphatic System, Respiratory System, Digestive System, Urinary System, and Reproductive System. Just as each system serves a unique role in maintaining human functioning, each system also reinforces and/or is reinforced by other systems. Planetary boundaries are not rigid; they are, like the human body, an ecosystem of systems and sub-systems that interact with each other. Creating an impact in one part of a system can result in an impact within and across other systems. Planetary systems, in the ecologic sense, are not separate from human systems, behaviors, or impacts. Rather, the vitality and functioning of human systems are predicated on the vitality of planetary systems. Human health and prosperity are intrinsically integrated with the health of the planet. Although this has been well-substantiated and is even a spiritual foundation to many world cultures, most of the human-built systems over the past 300 years have disregarded the sanctity of this relationship. Human survival and attainment of a greater quality of life are tied to the planet's capacity to provide resources, as it also carries out its self-regulating and reinforcing ecosystem services.

- Consumption is the beating heart, societal values and norms represent the neurons that connect cognitive function, and money represents the blood that circulates throughout our society. The world is much more complex than this, of course! But, when we contextualize sustainability and make it accessible to everyday citizens, we have a greater likelihood of helping people understand the immediate connection between their choices, decisions, and behaviors and the impact that is then put into motion through planetary systems.
- Sustainability is not the sole responsibility of any one country, company, industry, government, or inter-governmental coalition. Ultimately, sustainable choices reside with the everyday action or inaction of billions of people, at an individual and community level.
- While sustainability can be complex, it can be simplified so that clear win-win solutions can be attained. At the heart of demystifying sustainability is understanding people through their own lived experience. The planet is teeming with a rich diversity of life, including the diversity of different peoples that comprise humanity.

When the Teacher is Taught by the Students, *The Sustainability Generation*

Teaching is a privilege. Since 2018, I have served as an adjunct faculty member at the Whitman School of Management at Syracuse

University, teaching undergraduate and graduate courses in sustainable enterprise and management. Over the years, I've had the pleasure of meeting and teaching hundreds of students. I've also had the opportunity to coach and mentor students after their graduation, and in a few cases, guide and support their decision to pursue graduate or PhD degrees or obtain full-time jobs in the sustainability field. I'm grateful for having met all the students over my tenure. The art of teaching, in my experience, requires one to be adaptive, resilient, conversational, and improvisational. These are skills I've had to learn, modify, or adjust and relearn many times over. Each year brings in a fresh cohort of minds eager to learn, grow, and discover their place in the world.

One of the impetuses for writing this book is to honor the students I've taught and have gotten to know. Engaging with students week in and week out can be a humbling experience. It has required me to remain current, keep my mind sharp, and bring my best self, including my knowledge and experience, professional network, and new ideas to the classroom each week. The time and effort to do this well cannot be understated. But preparation is not even the hard part. Translating ideas and knowledge each semester into clear learning objectives that also resonate with each successive generation of students, each growing more worldly, wise, and discerning — now that presents a challenge to me as a practitioner.

Teaching sustainability in a management school has had its perks and limitations. I do my best to provide students with a comprehensive and balanced view of sustainability and the implications on society and the economy, and the imperative for business to lead. I try to incorporate perspectives in science, engineering, communications, and other disciplines and foundations. As a business school class, however, our lens is focused on creating sustainable value from enterprise strategy, management, development, and entrepreneurial pursuits. The business and management lens for creating sustainable (economic) value has its strengths, but also limitations when considering the vast interconnected complexity of global sustainability challenges and their root causes.

Year-over-year, a common refrain I would hear from students was "we can't consume ourselves to a sustainable future." Making businesses more sustainable, including their operations and capacity to innovate sustainable solutions for consumers, can significantly advance society's sustainability agenda. But solving sustainability through the power and influence of business alone will not be enough. If we truly want to live in harmony with each other and the planet, ultimately, the world's 8 billion people, who also comprise civil society, need to reconcile our role as global consumers and as citizens within planetary boundaries.

In 2014, when I published my second book, *Time to Trust: Mobilizing Humanity for a Sustainable Future,* I referenced that it was (and remains) convenient and easy for people to point their finger at Big Business, Big Brother (the government), Big Tech, and any other large institutions as the underlying crux of the world's problems, including those that flow to individual citizens and consumers.

While there is truth to the entrenched institutional barriers that reject change and incentivize a status-quo mindset among masses, it is important for consumers and citizens to be reminded of the enormous power we have. Through our pocketbook, our vote, our voice, and our influence, we have in the past, and can continue to change the world now and in the future, when the systems and institutions that govern, protect, and serve us are no longer working.

Our "Big-X" (X being media, business, government, technology, etc.) global economy is broken, it has been for a long time. There are big global problems to solve, and most know that the same "Big" systems that got society into our current state of global entanglements are likely not the governing bodies that can effectively aid society in a sustainable transition going forward. Big systems are frequently doomed to fail if they don't evolve, adapt, innovate, and change. Even institutions such as the Conference of the Parties (COP) have become mired in their process, a signal that perhaps their form and function may no longer be serving their full intention.

The COP is the main decision-making body of the United Nations Framework Convention on Climate Change (UNFCCC) conference, which focuses on global climate. The COP includes representatives of all the countries that are signatories (or Parties) to the UNFCCC. As of 2023, twenty-eight COPs have convened to discuss and prioritize global climate risks and concerns, and to direct resources to help ameliorate and expedite a more cohesive climate response. In recent years however, the COP process and convention has received scrutiny from climate activists, much as the annual World Economic Forum, held in Davos, Switzerland, has, for bringing together global elite from far reaches of the globe, who travel by planes, trains, and automobiles, consuming significant amounts of petrochemicals, further exacerbating the underlying top agenda item, namely carbon and the other greenhouse gas emissions.

Clover Hogan, a climate activist and founding Executive Director of Force of Nature, a youth non-profit organization that turns climate anxiety into action, characterized this juxtaposition in the following April 2024 LinkedIn post[20]:

> *"There were 2,456 oil and gas lobbyists at COP28. How many will we allow at #COP29? We won't solve the climate crisis with the same thinking that created it; or by relying on processes that have been so deeply infiltrated by people who benefit from business-as-usual. It sounds obvious, but we need decision-making structures that centre the people who are most affected: young people, people seeking asylum, women, BIPOC communities. And not only because these groups are vulnerable, or because they're victimised by the current system, but because they are the best placed to design inclusive and intersectional solutions. They have lived experience of being failed by the system, and they know what needs to change. We urgently need their perspectives, their wisdom, and their ingenuity."*
>
> *~ Clover Hogan, Climate Activist, Speaking at the European Union Agency for Fundamental Rights in Vienna in March 2024, posted and accessed on LinkedIn, April 4, 2024*

Over the course of my career, I've worked with a diversity of professionals. It's unfortunate that too many people are subjugated by the opinions or judgments of others, into "typology boxes" where they are reduced to personifications and characterizations of their work, as opposed to the greater sense of wisdom, inspiration, and constructive change that they profess and can manifest. In my view, Clover Hogan is not an activist, rather, a pragmatist who understands that the world is in a state of severe stagnation, brought about by unprecedented global group think, that has poked at, prodded, and twisted terms like sustainability, resilience, decarbonization, environmental justice, and climate action to suit their own specific needs and agenda.

Pragmatism provides the means to challenge and pressure-test truth — whether that is in speaking truth to power (activism) or lifting the veil of manufactured obfuscations that reveal the real hidden truths, the ones whose stories have been subdued, buried, and forgotten. It takes courage, character, and faith to be a pragmatist. I'm personally in favor of the Clover Hogans of the world. Their work is necessary to remind us that setting a direction is not enough — we also must not fall asleep at the wheel as we attempt to move forward.

The hundreds of students I've served are what I wholeheartedly believe represent *The Sustainability Generation.* Not to put a label on them (as I've found the younger generation does not appreciate this), but our youth are the ones inheriting our economic and social systems, and the planet in its current state. Think of this moment in society, and with the interconnected global challenges laid out by the United Nations' Sustainable Development Goals (SDGs) and a myriad of other human atrocities as a backdrop, as a home for sale, rent, or being transferred in ownership to family heirs.

Prior generations have done what they can to clean up the yard, fix the plumbing, paint the siding, and install a few new appliances — but for all intents, the house is being sold or transferred, "as-is," and the seller wants a quick sale with no inspection or other contingencies. Oh, and interest rates are extremely high! This is why I think of my students and most everyone younger than I as the Sustainability Generation. They were born into a world and required to adopt and adapt something they did not have as significant a hand in creating.

And they're now being asked to take over the rent, the lease, and the mortgage at a premium price in a market that is bearish (economically) and volatile (climate risk). Their main objective, as I see it, is to improve the home's foundation, reconstruct what's broken, and redefine what it means to be a homeowner on this unique planet. Prosperity, like democracy, is never guaranteed. It must be intentionally pursued and earned. Further, one's prosperity is strongest when everyone's potential to pursue and attain prosperity is equally strong.

As visualized by the image below, the foundation is flanked by four essential pillars: a physical (the Earth) pillar, a metaphysical (spiritual) pillar, a temporal (oriented to account for time) pillar, and a sentient (encompassing of life and all living things) pillar.

Pillars of Planet Pragmatism

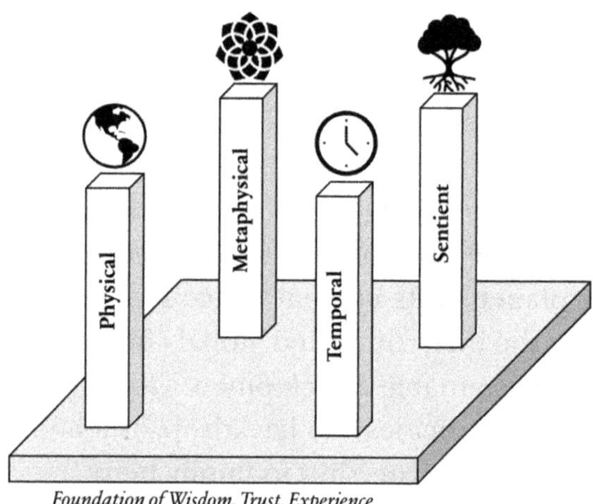

Foundation of Wisdom, Trust, Experience

The pillars of planet pragmatism are about examining and optimizing our coexistence with each other, with all living things, and how we continue to survive and thrive on the third rock from the Sun. At this point in our evolution, humanity has been built upon a foundation of wisdom, trust, and experience. The pursuit and attainment of prosperity requires us to establish a plan (call it a playbook) to guarantee our future. Our future can, however, only be guaranteed if we have the humility and patience to objectively learn from the past. Further, we must elevate the wisdom garnered across the four pillars of planet pragmatism and put that to beneficial use for the whole of society. Meaning, at some point, we must act. We cannot allow paralysis by analysis to keep humanity from moving forward, together. This is not to say we should make hasty decisions. Rather, planet pragmatism provides, as will be discussed throughout this book, a means for us to integrate wisdom, knowledge, and intention in a manner that envelops and integrates the four pillars of pragmatism. Over the millennia, our home has been remodeled again and again. It's now time to revisit our foundation so that we and future generations can continue to have a viable place to call home.

Points on Pragmatism

- *Sustainability is dynamic, always shifting with nature, and with the nature of society; Currently, we are not sustainable.*
- *Sustainability has become overly politicized, and it is subjective, so Start with Why.*
- *Our future can only be guaranteed if we have the humility and patience to objectively learn from the past.*
- *Four pillars of planet pragmatism include a physical (the Earth) pillar, a metaphysical (spiritual) pillar, a temporal (oriented to account for time) pillar, and a sentient (encompassing of life and all living things) pillar.*
- *Planet pragmatism provides a means for humanity to integrate wisdom, knowledge, and intention in a manner that envelops and integrates the four pillars of pragmatism. We must elevate the wisdom garnered across the four pillars of planet pragmatism and put that to beneficial use for the whole of society.*

4

IS SUSTAINABILITY DEAD?

Is the "S-word" Dead?

Like many other sustainability strategists, practitioners, researchers, and educators, I think about sustainability a lot. In fact, it is probably too much. I imagine that people in other occupations spend a great deal of their time deep in thought about their interests as well. Deep immersion in sustainability has its intrigue, but it can also be fleeting and frustrating — and sometimes, downright frightening.

The "S-word," sustainability, has its share of misguided misgivings. I deliberately say the S-word, because sustainability can be, if spoken about too broadly, ambiguous and confusing. This has been a detriment to moving [sustainable] action forward. Today, the S-word encapsulates a large spectrum of meaning and intent. A great deal of S-word focus is placed on evaluating climate risk, decarbonizing industrial sectors, gathering data to support corporate ESG reports and assurance, examining circularity within supply chains, and supporting "sustainable consumption" including in fashion, food, and even finance. The inclusion of more descriptive words and phrases works both for and against the S-word. In some cases, clearer language serves to create action and drive accountability; in other cases, new words and phrases can subjugate the deeper intent of sustainability.

The latter is an issue that I frequently think about. The expanse of S-word language is both helping and hindering the broader sustainability movement. If we are not careful, we may find ourselves fixing one element of sustainability, but at the expense of other and perhaps more critical factors that create impact on our health, safety, security, and the environment.

Many years ago, following the publication of my first book, *The Sustainability Generation: The Politics of Change and Why Personal Accountability is Essential NOW!*, I spent a fair amount of time speaking at trade conferences, colleges and universities, public events, and open book discussions. I enjoyed getting out, promoting the book, and opening thoughtful conversations on the topic of sustainability. I found that the more I spoke before non-technical audiences that had less knowledge about sustainability, the more interesting and often nuanced the conversations would be.

This observation became more apparent when, a few weeks into the S-word conference circuit, I realized that I was speaking to the wrong audience. When you attend a conference and "press the flesh," that is to say, shake hands with a bunch of S-word converts, you are among friends. At least mostly friendlies. I say mostly because, as with any professional network, S-word folk are competitive. Yes, they like to share knowledge and are typically service-oriented. But as human nature would have it, some S-word conference goers like to prove their smarts, often looking for those "gotcha moments" to correct someone, or let their peers know how well read, published, or credentialed they are.

Well, all that's nice, but it doesn't make the world any more sustainable, at least in my experience. Now, I'm not knocking the value of conferences or the seriousness and impact of S-word professionals. I'm simply making the point that if you are only out there speaking to those who are already true believers and practitioners, then your potential for impact is marginalized to that community of professionals. Attending conferences and meetings among like-minded and similarly experienced and trained professionals serves a valuable purpose. It is important for peers to get together to exchange knowledge, share best practices, and support each other within their chosen profession.

For the past few years, I've sensed that S-word change agents have inadvertently done ourselves, the profession, and the intention behind the philosophical construct a disservice. For sustainability to continue to advance, we also must be willing to listen, learn, and grow. We must also challenge our own mental models, be willing to engage with audiences we are not accustomed to or comfortable with, and help coach, train, and mentor those who are just learning about the ambiguous and multifaceted S-word. Too often, S-word conferences represent an echo chamber of similar points of view, experiences, and ideas. While this can and does contribute to a necessary convergence of ideas, the conversations and subsequent change can feel a bit incremental. Yet, the climate risks and planetary changes that a majority of respected scientists and experts are projecting signals to a swifter societal call to action than what comes out of S-word conferences.

Essentially, we can analyze and discuss all the data-driven, highly detailed and glossy, voluntary, prescribed, and regulatory-driven reports on ESG and sustainability all that we want, but that does not guarantee us a more sustainable future. Sustainability is as much, if not more, about what we conceive, design, and innovate — as it is about what we are trying to measure, modify, and fix. We cannot mandate, dictate, delegate, or regulate our way to greater sustainability. Sure, each of these tactics serve a role within the management of a sustainable marketplace, but they are only as useful as the market is willing to participate. The objective for a sustainable future runs deeper and wider than a ban on plastic shopping bags or coffee straws. Command and control, voluntary do-goodism, or incentive and market-based nudges are all tools for managing sustainability within existing fixed systems. This is important and necessary work.

For our capitalistic society to make significant progress, ultimately, we need to redefine the self-limiting systems that we've constructed and maintained. We must also reconcile our purpose and nature of being with the laws of the natural world and the Universe. Most, if not all, of the systems that humanity has built were not designed with an ethos of sustainability, let alone financial equity, inclusivity, circularity, or regenerative ecology and economics. In essence, we are spending our precious time patching the outdated and inefficient systems we allow ourselves to be beholden to as opposed to challenging this entrenched status quo to design a sustainable economy from the ground up. From where the S-word sits today, we have a long, long way to go.

A Nightmare on Sustainability Street

As a practitioner of sustainability, I am gratified to witness the mass market advance of "everything sustainable" that is underway. Twenty-five years ago, there were only a handful of academic programs or credentials focused on fostering expertise in sustainability. Today, there are hundreds of degree and credential offerings. A quarter century ago, S-word whisperers were trailblazing new terrain inside global corporations, working both behind the scenes and publicly to advance the "business case for sustainability," particularly among executive leadership.

Today, corporate boards are proactively asking difficult business sustainability questions and calling for more action. Subsequently, corporate executive leaders have begun to reinforce a culture of sustainability. Today, it is much more commonplace for corporate leaders to pull in, rather than push away, sustainability. Corporations are now much more fluid in how they bridge business strategy with the implementation of practical sustainability solutions that achieve business performance goals in step with stakeholder expectations and societal impact.

But for all the progress made within the global business community, there is something scary lurking around the sweet and friendly façade of sustainability street. A fixation on solving sustainability issues within the business, or within the industrial sector, is limiting deeper and more meaningful sustainability gains. Inherently, it makes logical sense for a singular business (and industry sector) to account for and address its sustainability risks and concerns. This approach can, however, lead to myopic self-interest for individual companies, industrial actors, and entire sectors.

The meteoric rise of voluntary environmental, social, and governance (ESG) disclosures has reinforced a competitive external reporting culture among global corporations. In the corporate reputation "report out or be reported on" contest culture surrounding ESG performance, companies tend to skew as protectionists first, and catalytic partners in solving cross-sector problems, second. There are many exceptions to what I'm describing here (i.e., corporate disclosures for double materiality[21], Scope 2 and 3 emissions accounting, supply chain agreements, public-private partnerships, etc.). Suffice it to say that when it comes to the current situation with sustainability, most companies are, first and foremost, incentivized to help themselves. For many S-word practitioners, this is understood, but it also runs counter to the holistic and systems-level mindset that we intuitively know is needed to create sustainable change.

Corporate reputation rules the S-word roost for the moment. The ESG and corporate sustainability disclosure movement is, perhaps, a necessary evolution to get all companies speaking a similar S-word language and accounting for their individual impact. But while we get our ducks in a row, the planet continues to roil its climate risks upon our doorsteps, much like Freddy Krueger callously thrashing his metal glove in our face. We've created our own nightmare on Sustainability Street. The planet is beckoning us to change or be changed. The "S-word" has in recent years become part of our sociocultural experience. The "S-word" has infiltrated our classrooms, our homes, our businesses, and our front porch conversations. The motivation for mass market sustainability is now more rooted in our collective psyche. The question now is whether the mass market sustainability of everything is sustainable. Have we shifted our thinking and behavior? Our perception of a mass market sustainability culture is not (yet) equating to a more sustainable planet as evidenced by the continued assault of climate risks and declines in biodiversity and ecosystem services.

Keeping "S-Word" Transitions from being Transgressions

Every industrial sector is currently pursuing a transition that is predicated on sustainable enterprise in some way, shape, or form. There are several meta-dimensions of sustainability, including decarbonization, decentralization, digitization, deregulation, and others that are working individually and together, as forces driving an economy-wide transition toward sustainable production and consumption. The intent of industrial sector-based and economy-wide transition is certainly necessary and noble.

There is a famous Albert Einstein adage that states, "We can't solve problems by using the same kind of thinking we used when we created them." In his book, *The Fifth Discipline,* systems scientist and MIT Sloan School of Management lecturer Peter Senge laid out eleven laws of systems thinking. The first of Peter Senge's eleven laws of systems thinking puts another perspective on this, stating, "today's problems come from yesterday's solutions." To ensure that we don't repeat the mistakes of the past, we must envelop a preventive, predictive, and proactive posture toward "planet prosperity." This is to say that we must leverage our wisdom and work collaboratively with common sense for the common good.

The current enthusiasm for industrial sector-based and economy-wide sustainable transition can quickly turn into unsustainable transgressions if we don't adequately listen, learn, and lead. To alleviate unsustainable transgressions, we need to intentionally bring together and listen to a diversity of stakeholders with dignity, and in an inclusive way. We also need to learn from each other, and our past, in a way that doesn't chastise or place blame, but rather enables us to discover wisdom and grow. Finally, we need to embrace an ethos of the "S-word" toward "planet prosperity" and ensure that all people have equal opportunity to lead, with courage, conviction, and resolve.

We Must Temper Our Desire to Control, Which is Out of Control

The natural world will not change to serve humans. Rather, we need to change our [human] nature in how we interact with and respect the natural world. For much of our modern existence, humans have attempted to exercise control over nature. Our desire to control nature has been fueled by our need for survival and our desire for comfort. The planet, while teeming with life, can be an unforgiving place. To survive and thrive, humans have had to both work with and around the natural world.

However, the scale of our cultivation and curation of resources has, as we know, created environmental damages and externalities that run counter to our need for survival and desire for comfort. Our more recent embrace of everything S-word has accelerated, in part, due to our growing recognition that our attempts to "nip and tuck" the earth into a place that serves us have come at an enormous expense to natural systems. What the planet once provided for free through ecosystem services (i.e., clean air, water, provisions of natural foods, climate regulation, etc.) we are now engineering "solutions" toward — not because the planet cannot provide these services, but because the planet's capacity to provide these services has been significantly diminished by our hubris in attempting to control nature.

So, we are now desperately trying to engineer ways to strip CO2 out of the atmosphere, prevent toxic algal blooms in freshwater reservoirs, predict and reduce the impact of wildfires, and prevent microplastics from entering our environment. The bill has come due to the price of our and prior generations' comfort. We now must pay the price by living with a swiftly changing climate — therefore succumbing, likely, to greater human discomfort. Further, we must pay the price, if we are to continue to survive and thrive, by investing in entirely new ways to attain comfort in a rapidly changing world. This requires us to evolve our thinking in how we work with nature, not against it, toward the pursuit of 'planet prosperity.'

To achieve planet prosperity, we must change our nature to restore and better protect nature, which nourishes us. We need to instill more common sense for the common good and elevate our intelligence and wisdom that provides a preventive, predictive, proactive posture on how we address challenges with each other, the role of technology in society, and how we can be a trusted steward of the natural world.

Don't just manage change, catalyze it!

Another one of Senge's famed eleven laws states, *"The harder you push, the harder the system pushes back."* Humans are stubborn, and we tend to trudge through difficult situations as opposed *to* accepting things as they are and seeking to reevaluate our best options. Senge refers to this phenomenon and our behavioral psychology as "compensating feedback." The more effort one exerts to improve or fix the system (i.e., sustainability within existing systems), the more effort seems to be required. For the "S-word" gurus out there, sound familiar?

For example, although energy efficiency has increased significantly over the past half century, energy demand has also continued to grow. Renewable energy now represents about one-third of the power generation creating electricity worldwide. That is an important accomplishment toward decarbonizing electricity production. Expand out of power generation, and the world's dependence upon fossil-based fuels holds steady at eighty percent. Further, in October 2024, the World Meteorological Organization (WMO) reported[22] that atmospheric carbon dioxide (CO2) concentration has increased by more than 10% in two decades, stimulated primarily by human combustion of fossil fuels. The WMO report found that CO2 is accumulating faster than at any time in human history. The WMO finding is interestingly and eerily in step with Senge's law. It appears, the harder we try to push solutions onto the existing system, the tougher the system is pushing back.

And yet, positive change is afoot. Demand for oil is projected to decline sharply[23] by the end of this decade, catalyzed by transportation and building electrification policies, strategies, and technologies. The energy sector, and the transition that is underway across all major industrial sectors and end-uses that it serves, illustrate both the need to manage and catalyze sustainable change. Society's requirements for energy are so entrenched with the form-factor of how energy is produced and delivered that it is difficult to simply turn the spigot off. Due to society's heavy reliance upon safe, reliable, secure, and now, sustainable energy, we need to be thoughtful and deliberate in how we turn one spigot off and turn a new transformer on.

Amidst our consternation and tension associated with the complicated energy and broader industrial and societal transition that is underway, there are some unique change catalysts that are set on reinventing the foundation by which humanity envelops sustainability in all that we are and do. Dubai/UAE, NEOM, and Toyota Woven City represent three examples of how bold sustainable change catalysts are working at enormous scale and dizzying speed to break down silos, disrupt existing systems, and create an entirely new society by reinforcing their future foundation with sustainable infrastructure, tourism, and living in mind.

The City of Dubai within the United Arab Emirates (UAE) has been an example of sustainability in action. Dubai, now popularized by modern architecture and luxury shopping, has been thoughtfully transitioning its economy for several years now. Although oil and gas exports continue to represent about 30% of Dubai's gross domestic product (GDP), industries including tourism, trade, financial services, and real estate have become prominent contributors to the city's economy. To achieve its goals for growth, Dubai has had to encompass many facets of sustainable development into the building out of the city's infrastructure. Dubai is pursuing carbon-neutral growth by enveloping sustainability into its master planning across all major systems (i.e., energy, transportation, water, food, entertainment/recreation). For example, the Dubai Electric and Water Authority (DEWA), which produces the majority of the city's electricity and drinking water, burns natural gas from Abu Dhabi and Qatar to generate electricity. DEWA captures waste heat from power generation to distill seawater, removing salt, so that it is potable. DEWA's operations produce up to 10 gigawatts of electricity and distill half a billion gallons of seawater daily.[24] DEWA has also invested more than $13 billion in the design and construction of the Mohammed Bin Rashid Al Maktoum Solar Park, a 5,000 megawatt (MW) solar park located in the desert outside of the city. The park can power as many as 1.3 million homes while reducing carbon emissions by 6.4 million tonnes annually.[25]

NEOM[26] is one of the five Saudi Vision 2030 megaprojects that fall within what is called Red Sea Global[27]. Red Sea Global aims to diversify the Saudi economy away from fossil fuels, including by expanding its tourism sector with a focus on sustainable innovation and development. It is estimated that as much as $20 billion will be allocated to complete the Red Sea Global project portfolio[28] by 2030. Once completed, the project will fulfill a tourism-focused masterplan that reaches over 28,000 square kilometers.

Toyota Woven City is another example of a large-scale development catalyzing human ingenuity toward greater prosperity. Described as a "living laboratory,"[29] the Toyota Woven City is exploring rapid adoption of green energy technologies and AI to initiate the construction of the city of the future. In Toyota Woven City, there is a focus on the future of mobility and "creating well-being for all."[29] Toyota Woven City has, for example, created an initiative called the Woven Test Course, where people co-create new ideas to make a positive difference in the world. According to Toyota, the cost of Woven City is over $10 billion, which will include everything needed for a functional city, including residences, stores, plazas, entertainment, and other amenities.

Dubai, NEOM, and Toyota Woven City are real-world mega-development projects with mega-goals that are stretching our imaginations and the limits of our existing technology and know-how. These cities and regions have their fair share of opposition from those who do not see these developments as sustainable or as pillars of greater prosperity. This said, in a world that is too often caught up in overanalyzing and managing change, Dubai, NEOM, and Toyota Woven City represent bold visions that are catalyzing action toward proactively creating a future for their people and culture. Each development has a unique sustainability signature that is worthy of ongoing observation and learning. The pursuit of planet prosperity and a more sustainable future is not linear or prescribed. It is and shall remain a bit messy and convoluted as we work to dismantle old, tired systems and introduce new ways to not only survive, but to thrive, in step with natural systems. Sustainability is not and will not be perfect. But it certainly is not dead. In fact, it is alive and with a fervor, redefining what's possible as we pursue planet prosperity together.

Points on Pragmatism

- *Sustainability, or the "S-word," can be an ambiguous topic and challenging pursuit. Deconstructing the S-word into action-oriented goals, objectives, and results can drive greater awareness, understanding, and adoption of the S-word. Each generation has its own unique S-word objectives, grounded by the current events and challenges they face in the moment.*
- *The planet is beckoning us to change or be changed. The motivation for mass market sustainability is now more rooted in our collective psyche. The question now is whether the mass market "sustainability of everything" is sustainable. Have we shifted our thinking and behavior?*
- *To achieve "planet prosperity" we must change our nature to restore and better protect Earth's nature, which nourishes us. We need to instill more common sense for the common good and elevate our intelligence and wisdom that provides a preventive, predictive, proactive posture on how we address challenges with each other, the role of technology in society, and how we can be a trusted steward of the natural world.*

- *The pursuit of planet prosperity and a more sustainable future is not linear nor prescribed. It is and shall remain a bit messy and convoluted as work to dismantle old, tired systems and introduce new ways to not only survive, but to thrive, in step with natural systems.*
- *Sustainability is not and will not be perfect. But it certainly is not dead. In fact, it is alive and with a fervor, redefining what's possible as we pursue planet prosperity together.*

5

CORPORATE SUSTAINABILITY TODAY: LOOKING TO THE PAST TO PREVENT FUTURE ENVIRONMENTAL LIABILITY AND CALAMITY

Part 1. The Lens by Which We See the Changing World, and Ourselves Within It

We now live in a complex global society that requires all of us to serve a role to ensure that economic, environmental, or social injustices to people or planet do not persist or promulgate. There are long-trenched power dynamics between civil society, industry, government, academia, religion, and non-state actors at work that are both passively and proactively shaping our world.

Most of us are subservient to the global chess game that is being played as we take stock of our personal livelihoods and spheres of influence. But individually, and certainly collectively, we have power and can positively influence our present and future. I am 100% in support of individuals, companies, and governments taking full responsibility and accountability for any negative impact that results from their behavior, operations, or policies. **To evolve as a society and as a global economy that enables prosperity for all people, we must be willing to learn from the past, be pragmatic in the present, and protect the promise of our collective futures.**

Dwelling on the past is a fool's move in this global chess game, as the future is being won now, in the moment. Reflecting upon and learning from the past is a necessary and smart move for those seeking wisdom and insight to inform one's next move, let alone the next three moves forward. We can view the world as complex and left to masterful chess players — and feel disconnected and helpless in what is unfolding and shaping our reality and quality of life — or we can stop feeling victimized, jump in, and change the dynamic of the game. But player beware, the rules are ever evolving. Staying sharp and making smart moves is a survival tactic and necessary skill for success.

In my short industrious career, I've worked through many moments of personal turmoil self-examining my purpose and the structured form and function of my chosen field and to those whom I serve. When I was a young lad, wide-eyed and fresh out of graduate school, I had a very binary sense of the world. The corporations that polluted were bad. Politicians who made do on their promises must always be noble and good. I inherently understood that the forces at play in the world were more dynamic than the simplicity of good versus evil. But it was then, as it remains today, more convenient to oversimplify the world and its actors.

In preparation for my formal professional launch into the world, for my undergraduate education, I pursued and earned degrees in humanities and social sciences, environmental studies, and geography. For my graduate education, I earned my Master of Science in environmental management and policy. Some might argue that my early academic research and training "converted" me to think of corporations, capitalism, and consumerism as "bad." Surely, I had plenty of those 101 classes that sought to evangelize environmentalism and capture my and other young students' attention by contrasting the evils of the world bestowed by multinational corporations. But, my training also brought me deeper into stakeholder relations and communications, geology, economics, consumer behavior, entrepreneurship, marketing and manufacturing, and so much more. It became, at least to me, self-evident that each of us has a unique lens by which we view the world.

The lens we each have, at any moment of time, is encoded by our DNA and engineered by our experiences — how and where we grew up, where we went to school and what we studied, whether or not we were nurtured and loved, our family genetics, our quality of health, and so much more. Although the building blocks of what makes us each unique can be generalized as similar, there are an infinite number of data points that nudge each individual's lens to be skewed slightly differently from another's. Take me and my three sisters as an example. Each of us grew up with the same generalized inputs and experiences from a family perspective, yet we all have a different point of view on our past as well.

Over the course of my life and career, I have found that the aperture of the lens widens, as I take in more information and am shaped by the people, places, and experiences that comprise my personal journey. The journey is distinct, and the assimilation of data and information drawn from prior experience is also unique to each individual.

Over the past six years, I have taught undergraduate and graduate students at Syracuse University's Whitman School of Management. I've often used the "lens by which you see the world" as a metaphorical exercise to introduce self-evaluation, critical thinking, mindfulness, diversity, and inclusion into the students' awareness.

When we become more aware of the **"lens by which we see the world, and the lens by which we see ourselves operating in the world,"** we can begin to better understand what has shaped it or shielded it, thereby identifying any biases we have in how we take in, evaluate, and make sense of the world around us. And by the way, we all have biases in how our lens operates. The sooner we can identify biases in how we take in, interpret, and act on information — the more quickly we can move from playing connect the dots, to checkers, to chess.

There are many self-identifying pundits in our world today. Everyone seems to be an expert on something. The development and projection of expertise can be limiting, particularly if the lens by which it refines that expertise does not acknowledge its blind spots, inherent biases, or areas for further discovery. The metaphoric lens, for example, may be blurred, smudged, or it may have a filter on it. Too often, "experts" remain confined to their worldview, limiting beliefs, and the rigidness of their workflow on how they choose to analyze information. Those skill sets got them to the game, but they will only play and advance so far.

When I was in my 20s, I was intrigued by global business. Admittedly, I thought most large companies were bad, faceless organizations that pursued profit at any cost, including degrading the natural environment, mistreating employees, and causing unhealthy living conditions around the world. Today, many of the same activist organizations that fought twenty-five years ago continue to pursue the same game (and fight) against the same companies today. The game goes on, so to speak. Watchdog activists serve a role in our society, no doubt. As newer generations of pundits rise for attention, greater agency and autonomy, my hope is that they spend time learning from the past, remaining vigilant — but not violent in the present, and humbled by the fact that the future they desire is but one perspective in a world of eight billion and more lenses.

As I am near a milestone of three decades of professional service, my aperture to the world continues to widen. Some of the skills that earned me a career as an energy analyst and management consultant twenty-five years ago are being disrupted by a new generation of capabilities, including artificial intelligence (AI), data science, and social media engagement. This said, I believe I've cultivated new skills over the years that I never thought would be as critical as they are today. Creativity, critical thinking, agility and adaptation, objectivity and self-awareness, leadership growth and development — these and other skills — I have found, help one to grow, lead, and discover through change.

Our life's journey is ever-changing. The person and professional I was twenty-five years ago is much different than who I am today. I like to think I've held onto the good, shed the bad, and continue to remain open to my biases and limitations and opportunities for personal growth and improvement. To that end, much like our personal journeys and the stories we attach to them to create meaning, so too are the stories of global actors, governments, NGOs, corporations, nation-states, and others. These stakeholders are comprised of people, individuals who are also serving a narrative and exploring their lens on the world. They, too, are evolving, growing, and attempting to lead. There is nothing perfect about this process, but it is important to acknowledge that this is a process, and people can intersect and disrupt, or intersect and invigorate and empower the process at any moment in time.

I do not believe companies are inherently bad or evil. I have students, colleagues, and contemporaries who have and, likely will after reading this, continue to challenge me on this point. Perhaps this remains a blind spot and bias that I need to reassess. But as I see it, companies are made up of of people, and I have a hard time believing all people are bad and set out intentionally to wreak havoc in the world. I can think of a few very bad people and corporations; they were overtaken by greed, poor leadership, and inept management. Eventually, those with ill intent were found out, and the people were removed, and/or the companies failed to exist.

The idea that corporations are evil is a pervasive cultural stigma that has created a blind spot on how some business schools train next-generation talent, and on how young and mid-career professionals think about how they can work with (rather than against) corporations to create sustainable value and positive impact. Unfortunately, corporations are viewed as ugly giants or fire-breathing dragons that need to be destroyed in order for positive change to prevail. Planetary pragmatism, the idea that change is both radical and intentional, challenges the notion that positive change only happens when bad actors are removed.

Part 2. Making Change is Not Always About Slaying Dragons. Corporate Sustainability is an Evolutionary Tale of Innovation and Pragmatism. Look No Further than the "Current" Energy Shift

Our society navigates to and celebrates stories of triumph, courage, and transformation. Who doesn't like a good underdog or comeback story? In popular culture, such stories have been romanticized by the media and within books and movies. But these stories have also been handed down for generations, much like the story of David and Goliath. There is something enduring about taking down the big, bad giant or slaying the fire-breathing dragon.

Amid the economic, social, and environmental challenges society now faces, it's no wonder that multinational corporations represent Goliath and/or the dragon, needing to be taken down and slayed. The oil industry is largely considered the Goliath of our time. The products and derivatives of this industry are foundational to our global economy. The industry fuels our society and touches every aspect of our modern world, from transportation to housing, healthcare, telecommunications, finance, pharmaceuticals, defense, education, and infrastructure — the molecules of the oil industry have bonded with all facets of our daily lives. The industry powers our homes and cars, it's part of how we diagnose and treat diseases, and it is a lifeline for national defense and security.

Subsequently, the world's largest corporations (some of which control greater wealth than nation-states) are oil and gas. Further, the next round of the world's largest companies are either enablers, benefactors, or both of our oil-based economy, including aviation, defense, automotive, chemical, and even technology enterprises. It's no wonder that oil companies have long been and remain the target of social and environmental activists, many of whom clamor for attention by defacing historic statues, landmarks, works of art, or by attempting to disrupt corporate Board meetings, events, and even the daily minutiae of civil society. The interesting thing about modern corporations is that they control enormous resources: financial, technological, information, human talent, geopolitical, trade, and union, and so on.

One must be careful in any attempt to take down the giant or slay the dragon. *Slaying a big dragon may only lead to the creation of many smaller, but equally hungry dragons that will, given the dynamics and forces of the current world order, feast and grow up rather quickly.* If those baby dragons grow within the same culture and thinking of the big dragon, little will have changed; and worse, the social, economic, and environmental challenges we face may be exacerbated.

This is not to suggest that giants cannot fall, or dragons should not attempt to be slain. Rather, it's simply a point in acknowledging that changing the world doesn't always require an underdog or comeback narrative. Sometimes change happens over time. The giant dies off, naturally and methodically. However, for many people, this process is frustratingly slow.

The oil industry has an enormous target square on its back, and that target is only getting bigger. Fed up with environmental disasters and plagued by climate and health risks, society is questioning the trade-off between an economy fueled by oil versus one that is fueled by more sustainable options. Radical innovation in energy technology and infrastructure can intersect and disrupt the oil industry. There is a great deal of prior and existing knowledge, knowhow, intellectual property, and capability that the oil industry has that has, and I believe will continue to serve humanity, as an energy shift ensues.

Truly, all of industry and society are undergoing an energy shift, call it a transition in the next two decades, and a full-on transformation thirty years from now. In the near term, the transition will continue to place emphasis on the existing energy value chain. Demand for energy is rising worldwide[30].

Underlying the demand for electrons and molecules are classic market values, including the need to have affordable and reliable energy. In recent years, demand for cleaner, diversified, and distributed energy has also become a core value for many customers. In the long term, the energy transformation will be shaped by the eight meta-dimensions of sustainability, including the digitization, democratization, decentralization, and de-risking of energy supply, delivery, and utilization.

The clean energy shift will take time. Much like society's love of a rags-to-riches story, we tend to love elevating technology entrepreneurs and placing them on pedestals. But for all its enormous promise to elevate prosperity, the technology sector, much like the oil sector, has its challenges. Look no further than the prolific and contentious Elon Musk as an example of someone either beloved or loathed. Musk, an iconic entrepreneur, is hailed by some as a savior, and as an obstructionist by others. On one hand, his company Tesla is working hard to decarbonize society by disrupting the oil-based automotive industry; on the other hand, his SpaceX rockets have punched holes through the Earth's ionosphere[31].

I'm more than willing to bet that the technology industry has, can, and will continue to justify high-risk technology deployments as justifiable, no matter the unintended [security, social, ecologic, or economic] consequence, in the name of advancing humanity for the long-term. For some reason, many staunch clean energy and environmental advocates tend to dismiss certain entrepreneurs or technology companies from the environmental risk equation. The fixation on slaying the dragon can be distracting. If we are not careful, what we might just discover is that one problem (i.e., carbon elimination and removal) was simply traded for another, just as persistent and pervasive, challenge.

The advance of artificial intelligence (AI) and coupled with declining costs and enhanced efficacy for renewable energy, energy storage (across the spectrum: grid-tied battery storage, transportation electrification and batteries, consumer-focused energy storage), and other energy technologies (i.e., advanced solutions for scaling hydrogen, geothermal, small nuclear reactors, and more), is providing options for curtailing fossil-based energy.

At scale, these technologies pose great opportunity for reducing greenhouse gases (GHGs), yet they also pose a new generation of environmental, social, and economic risks, not that dissimilar to the fire-breathing dragons they are attempting to slay and displace. Renewables, batteries, sustainable fuels, advanced nuclear — these and other energy shift solutions also embody risk. For example, the supply chain concerns regarding "shifting the burden" regarding environmental damages and health and safety onto rare earth and battery mineral (lithium, nickel, cadmium, etc.) mining communities in developing countries have been well substantiated and documented.

The environmental risk and societal costs associated with the extraction, conversion, distribution, use, and retirement (recycle, reuse, remanufacture) of rare earth minerals that enable our digital society, and the advance of clean energy cannot be deemphasized. It is true that there have been studies evaluating the lifecycle costs and benefits of a petroleum-based energy society versus a clean energy society, typically favoring clean energy pathways.

Turning off the oil spigot and transforming our energy systems to alternatives cannot happen overnight. Our infrastructure has not been set up to absorb the cultural, economic, or cyber-physical shock waves of swift swings. For better or for worse, markets enjoy a little bit of certainty — call it at least a forecast for the week's weather ahead.

We are seeing this play out in the North American power market, where policies pushing for electrification and decarbonization are butting up against the "physics of power," and the capacity for the existing transmission and distribution infrastructure of the power grid to absorb renewable energy at scale, particularly in a manner that meets the regulatory and market requirements for affordable, reliable, high quality, and resilient electricity.

Of course there are a myriad of policy, legislative, technology, grid-integration, and business model scenarios and "fixes," that could be put into place in a measured way; but right now, there is a widening chasm between the idealistic views of a decarbonized and electrified future, and the engineered capacity of the existing system to transition (let along transform) in the time frame many policy makers have laid out in states like California and New York.

The energy shift, much like addressing corporate sustainability, is not [exclusively] a technological or business case problem; it is a lesson in pragmatism and people [leadership]. Again, this is not to suggest that decarbonization and electrification should not be pursued at scale. Rather, we must be realistic and pragmatic in how we work together, and across sectoral, geographic, and political boundaries, to ensure that we address the energy shift in a holistic way.

Part 3. Back to the Future: What We Can Learn from Environmental Liability "Mistakes of the Past" to Provide a Preventive, Predictive, and Proactive Posture for a Sustainable Future

Many of my colleagues and even friends and family don't know that, for fifteen years, I managed and facilitated a private peer-to-peer workshop that assembled 25 to 30 of the world's largest environmental polluters. Given the sensitive nature and confidentiality of these companies environmental liabilities, years ago I signed non-disclosure agreements (NDAs) with many of the individual companies, and also, with my main client, a global environmental, architecture and engineering company that hired my small firm to design, develop, manage, and facilitate this network on an annual basis. Although the term of those NDAs has long passed, I choose to remain objective and confidential in my disclosure.

To be clear, the companies engaged in this network represented some of the most toxic and litigious environmental liabilities in the world. Their portfolios included legacy environmental sites that spanned United States-based federally regulated Superfund hazardous waste sites to international environmental clean-ups in ecologically sensitive regions of the world. I took my oath of confidentiality very seriously, not in fear of my own personal liability, but because I quickly discovered, even as a young emerging professional, that the corporate environmental movement and the future of environmental protection, resilience and sustainability, were much more nuanced than my young self would ever have imagined.

The contaminated portfolio of this network would read like a horror novel to anyone familiar with environmental toxicology and the persistence and impact of long-chain chemistries and molecules in our environment, and upon our public health. The companies in this network represented a diverse and vast portfolio of environmental liabilities which included perfluorooctanoic acid (PFOA), perfluoroctane sulfonate (PFOS), polyfluoroalkyl substances (PFAS), polychlorinated biphenyls (PCBs), trichloroethylene (TCE), chlorinated biphenyls, dioxins, hydrocarbons and solvents, arsenic and other heavy metals, agriculture fertilizers and compounds including phosphorus and nitrogen, and unfortunately, much more. For anyone who may be unfamiliar with these contaminants and chemical compounds, think Love Canal[32], *A Civil Action*'[33], or PCBs in the Hudson[34] — three prominent cases that you likely have stumbled upon.

The companies in the network voluntarily participated and represented a diversity of industrial sectors ranging from integrated oil and gas, agriculture, chemical, defense, railroad, electric utility, consumer product, automotive, aerospace, semiconductor, and others. The companies voluntarily participated in this peer-to-peer network to engage with one another on emerging issues, exchange knowledge, share best practices on safety, operational performance, technology, engineering management, and more. Essentially, this cohort of cross-sector corporations met annually to benchmark environmental risk and performance against other companies and walk away with new ideas and opportunities to refine, improve, and advance their environmental remedial programs.

Collectively, the network's annual environmental reserve and annual spend for managing environmental liabilities were greater than $13 billion and $2.5 billion, respectively. The network also represented (back in 2018) a market capitalization of $2.5 trillion, $1.8 trillion in revenue, $115 billion in net income, and they employed over 2.1 million people worldwide. Those are significant numbers, even for 2018. I share this aggregated level of information to make the following observations:

- Corporate environmental remediation is an enormous business unto itself. The art, science, and management of environmental liabilities requires pragmatism and innovation.
    - Corporations managing significant, regulatory-defined environmental liabilities objectively want to extinguish those liabilities (i.e., "get them off the books") as quickly, cost-effectively, and with the greatest efficacy (certainty of remedial outcome) as possible.
    - Accordingly, corporations have invested heavily in their people, the environmental professional talent.
    - The financial carrying costs of environmental liabilities and their impact on the business are significant. Accordingly, corporations also invest wisely in new technology, innovative clean-up methods, data-and-science-backed decision analysis tools, and project engineering and management systems to ensure their environmental liabilities are cleaned up safely, efficiently, and according to the legal requirements and toxicological thresholds.
    - In many instances, corporations clean up remediation sites to a level that is cleaner than the background level of measured chemistries within adjacent parcels.
- Corporations managing complex liabilities have become a knowledge and innovation center. Given the complexity of

their task, these organizations have adopted a "preventive, predictive, and proactive" posture on environmental risk management, choosing to learn from the past so as to alleviate the future liabilities from being created.

- The know-how, knowledge base, and intellectual capital associated with environmental liability management and restoration, as it pertains to managing environmental risk and attempting to prevent future liabilities from existing enterprise operations and assets, cannot be understated.
- There is an underlying base of knowledge and capability that large global multinational corporations have with regard to managing environmental exposures, including how global enterprises can design, build, operate, and maintain, and decommission manufacturing, research and development, and other operational facilities around the world.

My observation, having worked within the environmental profession over the past twenty-five years, is that the stigma and caricature placed on corporate environmental professionals are often misguided and completely wrong. When I was a young professional, I too cynically and naively bought into the idea (reinforced by my youthful friends and colleagues) that corporate environment professionals were biased protectionists, doing the dirty work of polluting corporations, trying to keep secrets buried and any financial liabilities (or worse) at bay. Truthfully, I've met a handful of these types of people over the years, regrettably, they exist.

The interesting thing that I discovered early in my career, and which has been a lasting element over the past twenty-five years, is that environmental professionals care deeply about the work that they do. In fact, corporate environmental professionals are just as passionate about protecting and enhancing the environment as their peers serving regulatory, policy, advocacy, watchdog, or litigation roles are.

The corporate environmental professionals, the people behind the veil of corporate brands and lawyers, represent grandparents, parents, community leaders, local sheriffs and law enforcement, business owners, volunteer firemen and women, among many other hats they wear. These professionals work with environmental engineering and consulting professionals, regulators, lawyers, technology firms, construction management, safety professionals, local elected officials, and so many others who also share a strong affinity for protecting the environment, conserving natural resources, and enhancing quality of life within the communities in which they live and work.

But for most companies, and certainly for any company that takes its license to operate in society seriously, the environmental professionals working for them are incredibly passionate, dedicated, and have high integrity. They choose these traits — because they believe in being stewards of the environment — but they also understand environmental law and the repercussions before them, including imprisonment, if they do not uphold integrity, accountability, and responsibility in the work they perform. All environmental professionals understand this, believe me. This is not to say that bad actors don't exist; they do. And perhaps there is a dragon or giant out there that should be slain and fall.

It is important to acknowledge that the knowledge, resources, creativity, and intellect of existing enterprise holds tremendous value in shaping our future. The lessons of our storied and complex corporate environmental past, especially the awful ones, represent an opportunity for all stakeholders to learn from the past, as we work together to protect our present and prepare for a more prosperous future.

Points on Pragmatism

- *We can learn a great deal by being more aware of the "lens by which we see the world, and the lens by which we see ourselves operating in the world," by sharpening our skills in self-assessment, critical and creative thinking.*

- *Some of the world's largest corporations that manage the most significant environmental liabilities are also leading an advance toward a more sustainable future with a preventive, predictive, and proactive posture on environmental risk management and sustainable production in the 21st Century.*
- *Corporate Sustainability today has progressed beyond the romanticized allure of a David and Goliath narrative. Slaying big dragons will only lead to hungry baby dragons. This is not an underdog story or a comeback one driven [exclusively] by radical change. Rather, it's an evolutionary tale of the role of business in society, and the role of society in positively shaping and reinforcing shared goals for economic prosperity.*
- *If not appropriately assessed, attempting to create a more sustainable future by taking down giants and slaying dragons can be laden with unintentional global economic, environmental, and societal risk and concern. If we don't learn from past mistakes and mishaps, we are doomed to repeat them.*
- *Corporate and societal values and virtues for planetary pragmatism are increasingly at play, giving rise to greater certainty of business outcomes and environmental risk management. Managing change is a critical and in-demand skill set for environmental leaders, and a requirement for achieving personal and professional success.*
- *To extinguish environmental liabilities or address climate-facing greenhouse gas emissions with new technologies, there is concern that we are also propagating a new generation of environmental, economic, and social risks.*
- *The energy shift is indicative of the broader industrial and societal transition that is underway, seeking to redefine prosperity in terms of planet pragmatism, an approach to uplift the inherent wisdom we already have for survival while optimizing resources to attain a better quality of life here and now.*
- *Humanity is, in a sense, living with one foot in the past and one in the future. Too often, we forget to leverage our wisdom, garnered from prior experience and know-how, with our desire to shape the future, to*

inform our decisions in the present moment. This phenomenon limits our collective lens (view of the world and what's possible) to achieve a positive impact in the here and now.

- *When we integrate our wisdom and deep knowledge, much like corporate environmental and remediation professionals have, with real-world challenges today, we can make measured improvements for today, as we also consider the full potential of transitional and transformational technologies and new business models.*

6

THE SUSTAINABILITY BALANCING ACT: BEFORE CORPORATIONS CAN BE SUSTAINABLE, THEY NEED TO FIRST BE PRAGMATIC AND RESILIENT

Corporate management is increasingly bombarded from all directions for greater ESG disclosure and financial performance. Corporate Boards and leadership want to reduce costs, optimize resources, and yet, grow revenue by innovating before competitors. Customers want solutions but aren't always willing to pay for them. The value chain is price sensitive, yet they are also asking for solutions and progress to reduce Scope 3 emissions. The state of business is always animated. However, over the past two years, the industry has seen an unprecedented and even unrealistic convergence of maddening requirements.

Do more with less. Show positive ESG outcomes. Deliver products with fewer resources, less waste, and greater value. Oh, and don't forget to turn off the lights and take the trash out!

As corporations internalize and make sense of compounding and competing stakeholder demands that shape the future of the enterprise, like ESG and sustainability performance[35], they must also be sure to stay operationally resilient in the present. The duality of remaining vigilant and enterprising is not a core strength for most organizations. To pursue and achieve both ongoing operational excellence and drive new investment toward a sustainable future, companies must adopt a pragmatic and prudent posture to ensure that their vital signs are continually monitored and strong.

The popular airline safety statement comes to mind:

> *"...Any change in cabin pressure will also result in the aircraft's oxygen masks dropping from the overhead panels. If you're in the unfortunate position of being on a plane that is rapidly losing altitude, you should secure your oxygen mask before helping anyone else — including children."*

For most companies, sustainability can feel, and it certainly is, elusive. The definition of sustainability is open-ended and opaque. Like leadership, sustainability can be a bit subjective, requiring many different points of view to provide credence to how it is impacting the organization, customers, and broader society. And like leadership, sustainability requires periodic calibration garnered from continuous stakeholder engagement (from an organizational perspective), and from each subsequent generation of citizens and consumers (from a societal perspective), to remain evergreen and impactful.

Specific goals and targets under the corporate sustainability umbrella, like reducing carbon dioxide, can be abstract, particularly when framed as a climate change mitigation strategy. Of course, if all high-emitting global corporations attained their CO2 reduction targets by their established dates, that would be an enormous step toward mitigating future impacts of carbon dioxide. But how and when will we know if we truly abated enough carbon emissions to curtail climate catastrophe? Who makes that determination, and how? Is it grounded in a sustained measurement of ambient temperatures that fall within some threshold that humans accept? Or is it tied to a reduction in severe storm events or insurance payouts due to natural resource damages?

When sustainability goals and targets are abstract (i.e., reduce CO2 emissions by 30% by 2040), it is much more difficult for individuals and organizations to understand (let alone get excited by) the purpose (i.e., identifying with their "why"), to act and stay the course to attain the objective. Another critical challenge for organizations pursuing [abstract] sustainability is that too often, the most basic (yet critical and challenging) tenets of their operations can be put into question.

Sustainability Will Not Happen Without Immediate and Sustained Emphasis On Safety, Security, and Quality of Life

On February 3, 2023, at 8:55 pm, a 38 railcars from a freight train operated by Norfolk Southern Railway derailed in East Palestine[36], Ohio, releasing hydrogen chloride and phosgene into the air, forcing an evacuation of residents over a 1-mile radius. The East Palestine train derailment became an immediate public health and environmental crisis and emergency response incident.

It also stimulated intense immense stakeholder engagement from local citizens, politicians, and regulatory authorities and led to a focus on railway working conditions and safety including the lack of modern brake safety regulations, the need to implement precision scheduled railroading (PSR), and industry-wide questions around the length and weight of trains and the number of workers per operational train at any given time. Regulatory hearings, public meetings, and lawsuits ensued.

In April 2024, the Norfolk Southern Company agreed to pay $600 million to settle a class-action lawsuit, providing payment for personal injuries in a 10-mile radius and additional compensation to residents and businesses within a 20-mile radius who experienced disruption during and post the derailment event. Then, in May 2024, the company further agreed to a $310 million settlement for its role in the East Palestine derailment that contributed to public and environmental exposure to hazardous chemicals into the surrounding groundwater table and nearby waterways.

Leadership, Resilience, Sustainability, Safety ... These Values Are Everyone's Job in the New Economy

The pursuit and attainment of sustainability is not solely a corporate, government, or societal endeavor – it is all the above. Our modern, industrialized, and increasingly digitized society has a great deal of embedded and persistent environmental, public health, and economic risk. These are real operational issues that many companies, government agencies, and civil society must contend with daily.

Too often, people become complacent about the enormous number of resources, time and effort, and continuous monitoring, inspection, and improvement that takes place to literally keep our trains on the track and society running smoothly. When systems fail, we immediately see and feel the impact, whether it is a train derailment, an oil spill, a cyberattack or software glitch, a food recall, a bridge failure, an airline maintenance issue, or any host of other systems-level stressors that make headlines and are caused by human ignorance, poor judgment, or outright error.

How can our economy and society become more sustainable if the most essential functions of public infrastructure, like safety and security, fail us? How can we be more sustainable by rewarding profitability above

pragmatism? This sentiment is not directed at placing blame on any one company, industry, government agency, community, or individual. Incidents like East Palestine, the Deepwater Horizon tragedy and oil spill, or the 1989 Exxon Valdez Alaskan oil spill can certainly be prevented. It is also highly probable that society will be impacted by future incidents that negatively impact public health and the environment.

We must recognize and respect that securing a more prosperous present and future, including ensuring the safety and security of our private and public infrastructure, is never a guarantee. As citizens and consumers, we must all be willing to actively participate and serve a role in the continuous monitoring, feedback, and improvement of our built environment and society. Thus, if a situation seems unsafe, call it in. If you're being asked to do something that doesn't feel right, ask for clarity and seek additional perspective. Serving an active role as a citizen, employee, community member, etc., is a basic condition of how people should engage for a well-functioning democracy and society to work.

That sentiment, while grounded in pragmatism and a sensible "sustainable philosophy" and thinking, is not widely accepted. Sadly, we live, work, and play in a highly litigious society. Assigning blame, winning the verdict, and getting a big fat check for any damages is our modus operandi. Of course, the rule of law and establishment of legal rights and protections are a necessary foundation for any society. But when the system skews and shifts accountability away from the masses to only a select few, the overarching hierarchy of governance, democratic values, and ultimately control, becomes skewed as well.

In a hyper-litigious environment like we have in the United States, sustainability sounds great, but not if we're inconvenienced in any way, shape, or form. People don't want to go out of their way or be bothered with abstractions that don't have an immediate return for their lives. Symptoms of systems-level stress, such as inefficiency and waste, frequently make headlines, like airline delays, congested roadways, poor drinking water quality, computer software malfunctions, and so on. These are symptoms of stress within our infrastructure systems, and these events feel as if they are increasing in frequency, severity, and duration.

It is quite incredible how well our existing systems tolerate stressors. We take for granted that (for most regions of the U.S.), systems generally are working — energy, transportation, public health, financial, security, food, and so on. This said, we've taken some systems for granted, presuming they will always be there, delivering value and service as intended, and not anything we should ever concern ourselves with. That has created a false sense of security in our systems, something that is becoming more evident as systems are stressed and occasionally fail.

As our societal demographics change, our needs evolve, creating different and new demands for infrastructure services. Much of our existing physical infrastructure is aging and has not been adequately maintained, upgraded, or modernized. Simultaneously, new technologies are being deployed to provide new services, further rendering some elements of infrastructure inadequate or obsolete, which further puts financial pressure on continued maintenance. The net result of dynamic change and attempts to maintain the old while infusing the new is a continual flexing of the state of the system, which is increasingly more stressed.

Whatever our definition of prosperity may be, and whatever virtue and value we prioritize toward prosperity, be it safety, health, security, independence, clean air and water, education, economic opportunity, and so on, then we must be willing to get off the bench, roll up our sleeves, participate and work for it. Nobody and nothing are going to step in and hand prosperity to us. Democracy, capitalism, freedom, and independence — none of these ideals are guarantees of prosperity or privilege. I'd argue that the privilege that they do afford us is the right to continually pursue them, day in and day out, to ensure their longevity, as foundational pillars for how we choose to embrace life, liberty, and the pursuit of happiness.

At any given moment, the fragility of our human-built systems can be revealed. Human error is an element of risk that is infused throughout our modern economy and society; it's part of us. Although some people postulate that artificial intelligence (AI) can dramatically reduce human error, we must recognize that the building blocks of AI, in all its current and future forms, originally stem from human intellect, design, manufacturing, and interaction. AI needs fuel (energy), instructions (code), and governance (rules and structure). All of those requirements for AI to thrive are tied to infrastructure, systems, and structures that continue to require a human in the loop. This point may be self-evident; however, it is worth remembering that there is no panacea when it comes to fully removing human-induced errors or influences from the technologies, systems, and infrastructure that uphold our current economy. This reality should not hinder us from attempting to remain vigilant and ready to mitigate any risks that impede our health and safety, environmental quality, and capacity to prosper.

The Sustainability Balancing Act: Performance and Profit Through Pragmatism

Most companies are doing the "sustainability balancing act," akin to riding a unicycle across a high wire in a stiff headwind, juggling rare bird eggs, while onlookers throw rotten tomatoes. The direction of the high wire represents the future. The unicycle represents the existing enterprise. The stiff headwind signifies the uncertainty in the market. The rare bird eggs personify innovation and new products, the leadership team and employee talent mix, proprietary information, and partners. The onlookers include the diversity of stakeholders, investors, customers, regulators, suppliers and vendors, community organizations, and others, all waiting to take their shot at the unicycle.

The Corporate Sustainability Balancing Act

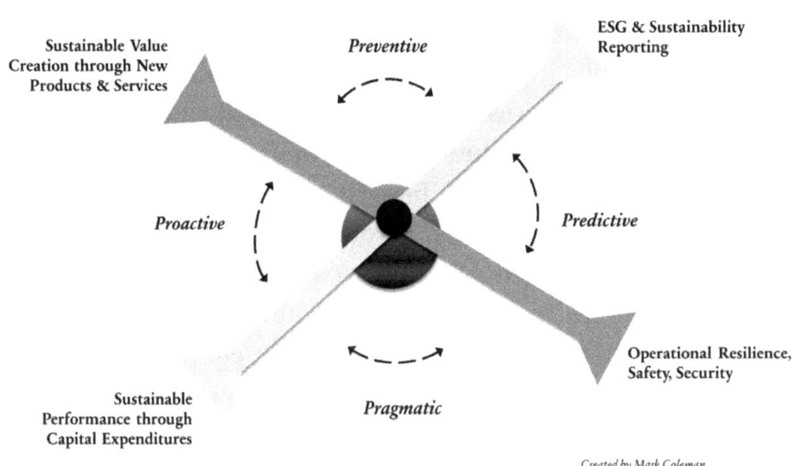

The current business sustainability environment is working hard to keep the unicycle steady and always moving forward, without dropping any of the eggs. Some core elements of the balancing act include:

- **Sustainability happens in the moment.** Corporate and social enterprises must seek to provide timely delivery on internal and external corporate risk, sustainability, and ESG reporting obligations. The corporate (and societal) pursuit of sustainability should not and cannot be an abstraction from addressing and delivering upon our immediate needs. If the ideals and investments in sustainability exist only in the ether, then we will continue to prod along and suffer the consequences of human errors in our products, systems, and infrastructure. The *sidebar* below delves into this in greater detail.
- **Identify, prioritize, and invest in meaningful operational and capital expenditures that seek to achieve operational and value-chain (i.e., Scope 1, 2, and 3) climate, waste, efficiency, and other ESG and sustainability targets.** Stakeholders want to see investment, action, and results. The days of simply quantifying the target and reporting out on incremental

improvements are limited. Institutional investors, NGOs, consumer organizations, and others are pushing companies to show their return-on-sustainability in a way that accomplishes financial objectives, and social and sustainable impact.

- **Remain competitive on innovation and new products/services that create value and deliver functional, social, emotional, and financial benefits through customer-facing "sustainable" solutions throughout the entire corporate value chain** (i.e., to suppliers, vendors, other systems providers such as energy, logistics, IT/digital, etc.).
- **Ensure the safe, resilient, and adaptive continuity of business operations, particularly as value chains, markets, and entire industries transition toward lower carbon energy, materials, and relationships.** As industry transitions to lower carbon inputs, the traditional values of environmental, health, and safety (EHS), quality, and efficiency cannot be forgotten or sidestepped to create and accelerate "sustainable change." There is compounding pressure on businesses to invest in the future However, they must also maintain and reinforce existing infrastructure and systems to ensure a safe and smooth transition.
- **Profitability (and attainment of greater prosperity) stems from planet and people pragmatism.** We must be willing to actively participate with our systems, our companies, and our economy to ensure bad things don't happen, and great things do. Corporations must continually assess and address operational risk and infuse sustainable thinking into their actions and behaviors in the present moment. In doing so, they will envelop a more pragmatic, preventive, predictive, and proactive posture in how they factor in, assess, respond to, and mitigate risk that directly or indirectly impacts their operations and value chain, and greater society's goals and priorities for greater prosperity (i.e., education, safety, security, public health, environmental quality, short-and-long term economic development and sustainability).

Real and Lasting Sustainability Happens in The Moment

Over the past five years, the rapid ascent of global sustainability disclosure and reporting frameworks has been astounding. The rising and turbulent tide of financial, accounting, regulatory, NGO, and trade-based standards has promulgated a new generation of business acronyms that seek greater data transparency, business (and executive) accountability, and enterprise impact. This alphabet soup of new reporting obligations has spelled trouble for global corporations who have been ratcheting up their regulatory-required and voluntary environmental, social, and governance (ESG), and more broadly, sustainability-focused disclosures to remain responsive to the growing expectations of salient shareholders and stakeholders.

Global companies have plunged headfirst (in some cases without flotation devices) into the stormy acronym sea to strategically posture among their peers as they compete for capital, reputational advantage, and social significance. The adage, "what gets reported gets measured," has provided credence to the advancement of the ESG and sustainability transparency movement. Truly, if there is a business metric that is monitored, measured, and reported upon, it must be important, right?

With ESG and sustainability disclosures amplifying the requirements and process of corporate reporting, the toll on corporate boards, governance, accounting and finance, investor relations, and ESG/sustainability teams has been significant. Recent emphasis on ESG reporting has increased the overhead associated with disclosure. Companies now spend more time and money both internally and with external counsel, accounting, and specialty consulting firms to ensure they can meet the rigorous timeline and nuanced requirements of all the acronyms riding the disclosure wave.

What's troubling and uncertain (call it an informed, professional "hunch") is how long companies can safely surf the disclosure wave until it smacks down against the sharp reef and rocky coast comprised of activist investors, citizen scientists, regulatory reformers, employees, and consumers. Transparency does not equate to sustainability, and the acronym soup of disclosure requirements has elevated the risk of a "lost in translation" moment for business sustainability. So that I'm not misunderstood here, let me say that I'm not against corporate transparency and disclosure, particularly as focused on sustainable enterprise and impact. The question that will soon require reconciliation, however, is whether or not individual companies, supply chains, business sectors, and the entire global economy are enabling greater prosperity, grounded in sustainable principles. Amid all of the ESG data analysis, disclosure, rulemaking, and polarization of perspectives — is the economy actually becoming more sustainable?

The Widening Chasm Between the Reality of Corporate Disclosure and Stakeholder Expectations

Unfortunately, the emphasis on ESG and sustainability reporting and disclosure has muddied the waters of corporate sustainability strategy, innovation, and impact. Reporting, while an element of contextualizing and understanding corporate sustainability, has served more as a proxy for "everything sustainable." Many sustainability practitioners understand the nuance, but many stakeholders do not. Anyone can do good on a test (particularly when the questions and ground rules are known in advance). A truer measure of [sustainable] performance is marked by intent, action, and impact, not just the careful preparation of curated data.

There is a large and widening chasm between the purpose, product, and functionality of ESG and sustainability reporting and disclosure and the idealistic hopes and expectations of the beachgoers watching the wave rise and accelerate toward the shore. Corporate transparency and disclosure play a big role in legitimizing sustainability. That said, sustainability is never a "one and done" metric, report, or financial filing. Business sustainability is dynamic and requires not only a retrospective accounting of investment and impact, but a pragmatic and principled stance on every singular moment. Retrospective reports cannot ascertain "in the moment" principled leadership, strategy, or intent. They can retrospectively provide a glimpse into corporate reasoning. Business happens in the moment, and our future sustainability is predicated on what business is doing right now and how it is envisioning its future.

Can the modern sustainability agenda bridge the widening chasm that exists between stakeholder expectations and the realities of ESG disclosure and reporting?

Real and Lasting Sustainability Happens in the Moment

Having a future-focused, long-term intergenerational perspective is a cornerstone of sustainable development. Meeting our needs in the present without limiting or hindering the ability of future generations to meet their own needs is the classic sustainable development definition. The natural resource choices and decisions we make today have a direct correlation to the capacity of future generations to meet their own needs.

We do not know what the future needs of society may be. We can make logical assumptions and projections on where the economy is heading and what the future may look like. For example, we can reasonably assume, based upon our current knowledge, data, and understanding of society, that the population will continue to grow, creating competition for natural resources. Advanced technology offers enormous potential for resource optimization and enhancing our quality of life, but only if we intentionally design and integrate technology with humility and within moral and ethical principles.

We cannot (yet) predict, with more than a few minutes or hours of notice, when a volcano may erupt; when a tornado may hit; or where, when, and how any number of planetary or societal extreme risks may spiral into something devastating. Thus, the future needs of society are confounded by generalizations and unknowns.

Change is omnipresent. While we like to try to predict change, managing it can be fleeting, and often, we fail. We cannot predict, at least with any great precision or certainty, the state of the economy or society in twenty, fifty, or a hundred years from now. We can only generalize. This should not deter us from trying to see into the future. Although our soothsayer skills are limited and we cannot foretell the future, we can influence it here and now. The actions, behaviors, and decisions we take and make today (i.e., political, social, economic, professional, just and ethical, etc.) directly impact the future.

Change is omnipresent. The future and fate of business and society reside in our capacity to accept, adapt, learn, grow, and lead through change. On a personal level, our decisions and activities in the moment impact the health and well-being of our mind and body. There is a delay in time between our decisions and the outcomes of those decisions. Sometimes the impact of our decision may be felt immediately, yet in other instances, the impact may not be seen or felt for a very long time. Famed MIT systems scientist, Peter Michael Senge[37][i], discussed the time and effect phenomena as one of his "11 Laws" in his book, *"The Fifth Discipline: The Art & Practice of the Learning Organization."* Senge noted in his Law 7, "Cause and effect are not closely related in time and space," meaning that we may not see positive or negative consequences of our actions for weeks, months, or years. Consider any bad habits such as smoking, drinking alcohol excessively, eating a diet of fatty and sugary foods, or having a sedentary lifestyle. All these behaviors may not have an immediate negative consequence on the mind and body, but if sustained over time, these choices can lead to serious health consequences, including diabetes, obesity, heart disease, cancer, and stroke. The decisions we make now, in the moment, are a portal to our future self and society. Making lots of good decisions can lead to an optimistic future. Making an abundance of poor decisions leads to a challenging future.

How an "In the Moment" Sustainability Posture Will Drive a New Prosperity

American (and human) prosperity is increasingly being called into question. The affordability of housing, food, education, healthcare, energy, transportation, and consumer goods and services are top and front-line concerns for consumers and voters. Further, entrenched and longstanding market failures perpetuate environmental, social, and economic issues, which continue to plague disadvantaged peoples and communities, further exacerbating equity and justice concerns. Topically and pragmatically, focused attention on "a new prosperity" gets to the beating heart of the pervasive divisiveness that continues to pit many Americans against each other and shape the unfolding political landscape and agenda.

While some politicians (and citizens) long for the prosperity of yesteryear (one, by the way, that is a romanticized relic of the past), others are focused on the seemingly endless firefight for a prosperity that has never really actualized for them. Mired in "the good ol' days" of what could have been of the past and overburdened and exhausted by the rhetoric of the present, most people simply don't have the energy or focus to pull together to formulate a new and shared prosperity. To do so would require unprecedented leadership, something which we are all starved of and desperately needing. The leadership we need needs to transcend the echo chambers of tech elite, political pundits, partisan politicians, and (un)social influencers. It's a tall task and big, likely thankless, job. But for those with courage, tenacity, and humility, there are multiple job openings that need this new kind of pragmatic and principled leader.

However, if we allow ourselves the opportunity, sustainability can provide the foundation for all people to positively and proactively shape our prosperity today, and well into the future. Americans (and all nations) are in a fight to adapt, survive, and thrive. A new prosperity is emerging, one defined by a deeper innate wisdom, a keener application of intellect, and a more dignified and humanistic value for all people and living things. We live, work, and play in an imperfect global society. The longing for a mythical version of prosperity that dates back one, two, three or more generations ago should not be our aim. Looking toward the past is a distraction from the realities of the present, and requirements of the future. The focus of our attention must be here, in the present moment. Our pursuit of prosperity should be measured, unified, and resolute. We can get there by having more constructive dialog, actively listening and learning about each other's needs and perspective, and shedding ego and irrational behaviors that prevent people from working together.

Sustainability is not just a corporate scorecard relegated to quarterly earnings, reputational jousting, and securing new investments. If conceived and deployed as a foundation for adapting to change, enhancing quality of life through innovation, and enabling a new generation of principled leadership to emerge — then sustainability will finally and unequivocally "be in the moment, and have its moment" toward achieving a new prosperity for all.

Corporate Sustainability is Not Dead. Like Prosperity, it is Evolving

No singular corporation is sustainable. No industrial sector is sustainable. And the power and might of sustainable business will not exclusively save people and planet. Hopefully, these statements are not all that shocking. For more than thirty years, a great deal of emphasis has been placed on corporate responsibility and sustainability. Most recently, voluntary corporate reporting and disclosure on environmental, social, and governance (ESG) has ballooned into a massive industry requiring expertise from accountants, lawyers, risk managers, and countless subject matter experts.

For the duration of my career, serious groupthink for business sustainability has been occurring. To be honest, I have, like many others, been drinking the corporate sustainability Kool-Aid. For much of the past three decades, "green business" and corporate sustainability evangelists fought hard with regulators, investors, executives, and other salient stakeholders to demonstrate that there is, in fact, a business case for sustainability.

Following years of practice in presenting their case, sustainability practitioners began to speak the language of business, finance, and management quite fluently. During this time, greater awareness and understanding of sustainability-related issues, risks, and opportunities were also trending across civil society. Business leaders recognized that a significant social shift was underway, driven by the collective changing behaviors, preferences, and values of critical stakeholders (i.e., citizens, consumers, politicians, regulators, investors, academic researchers, employees, NGOs, and others). Over the past three decades, corporations transitioned from needing to prove the business case for sustainability to defining sustainability goals, to implementing and integrating sustainable measures within the enterprise and across their value chain.

For the better part of 25 years, I have been working at the intersection of sustainability, enterprise development, education and entrepreneurship. I've worked for government, applied research and technology commercialization, business incubation, manufacturing, academic, engineering and management consulting organizations—all with a focus on clean energy, sustainable production, and innovation. I'm proud of my career, yet I cannot help but think that there is so much more work still to be accomplished. While tremendous progress has been made toward advancing sustainable production and products through the power and lens of business, our consumption-based society is eating itself, and we remain unsustainable. To achieve a more sustainable planet, responsibility and solution lie then, not only with business, but with us.

We can point our fingers at big business, big government, big tech, big agriculture, and big anything as the main culprits that perpetuate our unsustainable economy. However, the root cause of our deteriorating prosperity ultimately flows back to "we the people." There is also a litany of structural issues in our global society, ranging from market failures, status-quo thinking and policies, institutional barriers, and such that reinforce our unsustainable economy day in and day out. Placing blame on the old guard is just a means for us to skirt our individual and shared responsibility as citizens and consumers to further defer rethinking our best options to calibrate our behaviors toward creating planet prosperity. We cannot fix the past, and the future is what we make of it.

Sustainable enterprise is a necessary and noble pursuit in our Planet Pragmatists' Playbook. However, the creation of a sustainable global economy remains elusive. There is no doubt that corporations have a critical role to serve; and have enormous power and influence in how our economy transitions today and in years to follow. As the pendulum of pragmatism swings back and forth in the next five years, our capacity to develop and act on critical thinking, adaptive strategy, creativity, and innovation will be challenged, and yet it will be paramount to our pursuit and attainment of prosperity. The reins of responsibility cannot be passively ignored. The challenges we choose to prioritize and focus on, and the decisions we make today, will define our future prosperity. Therefore, we must grab the reins, hold ourselves and others accountable, and discover the power in pursuing a future that bridges common sense for the common good.

For business, this will require taking a critical evaluation of the efficacy of our modern corporate sustainability movement. We need to assess the overarching impact corporate sustainability has made toward the progression of industrial sectors, geographic regions, global supply chains, and the global economy, not just the myopic focus and attention that has been paid to singular corporate actors through ESG reporting thus far. There are, for example, companies including the retail giant Wal-Mart, shipping giant Maersk, and transportation innovator Tesla, that have each pursued deeper supply chain innovation that is yielding positive sustainability impact. The efforts of these companies and many more show that the opportunity for sustainable enterprise extends far beyond the proverbial corporate mansion and the products or services with which they align with consumers.

Business, like other sectors and stakeholders in our economy, is evolving. Traditionally, successful business enterprises create products and services that align with customer needs based upon clearly defined functional, emotional, and social benefits. This is referred to as achieving "product fit," and is an essential requirement in defining the value proposition of a business[38]. Planet pragmatism provides a means to prioritize customer needs in their journey toward greater prosperity. Accordingly, planet pragmatism explores the functional, social, and emotional needs of humanity. It also goes further to consider the temporal (time), cultural and spiritual context and needs of people and planet as well. We live and operate in an interconnected world. Our connections are temporal (across time), physical, digital, and spiritual. The power and influence of multinational corporations over these connections is significant. At stake is our sustainability and prosperity.

Corporate Sustainability, A Critical Tool Within the Planet Pragmatists' Playbook

A company that sells a sustainable product or provides a sustainable service does not guarantee that its enterprise is sustainable. Take, for example, Tesla Motors, Inc. (Tesla). Although Tesla is considered synonymous with Elon Musk for many people, it was incorporated on July 1, 2003, by Martin Eberhard and Marc Tarpenning, who served as Chief Executive Officer and Chief Financial Officer, respectively. Elon Musk was one of the early strategic investors in Tesla, however, he was not actively involved in day-to-day operations for several years. Musk did not become CEO of Tesla until October of 2008. Financed by tech entrepreneurs including Google co-founders Sergey Brin and Larry Page, and former eBay president, Jeff Skoll, Tesla also received hundreds of millions of dollars in loans from the U.S. Department of Energy, which the company repaid with interest.

Known for the idiosyncrasies of its iconic and controversial CEO as much as its products, Tesla designs, manufactures, and sells battery electric vehicles (EVs), stationary battery energy storage systems for home-to-grid-scale integration, solar panels and solar home shingles. In a relatively short time, Tesla grew from relative obscurity to one of the most disruptive and transformative automotive manufacturing companies in modern times. In 2024, Tesla[39] boasted a production of more than 1.7 million vehicles and 31.3 gigawatt hours (GWh) of battery energy storage systems, earning the company top-line revenue of $97.7 billion and a net income greater than $7 billion. Car sales contribute approximately 90% of Tesla's revenue, the remaining 10% is from energy generation and storage sales. Twenty years since its inception and Tesla is now the world's #1 electric vehicle manufacturer. As of 2024, in the United States, Tesla sells more EVs than Ford, Chevrolet, Hyundai, Kia, Audi, BMW, and Toyota combined. In fact, 55% of the EVs sold in the first quarter of 2024 were Tesla EVs[40]. But does Tesla's standing as the number one EV company in the world mean that it is the most sustainable? Certainly not. In fact, ESG rankings, including S&P Global, rate Tesla's ESG score below the industry average.[41]

There is no doubt that humanity can benefit from having alternative options for clean transportation and affordable clean energy. Curtailing the production and use of fossil fuels is an important step toward reducing greenhouse gas emissions as well as lessening water pollution and land degradation. Clean energy solutions, including transportation electrification, battery storage solutions, and decarbonized energy, are certainly part of our sustainability playbook. But these solutions alone are and will not be enough for all of humanity to thrive, not just survive.

From a corporate perspective, the definition, pursuit, and attainment of sustainability are subjective. From a societal perspective, there is no subjectivity when it comes to our ability to survive. We either do or do not. Right now, our planetary limits are being exceeded by unfettered, unsustainable production and consumption. We cannot buy our way to a more sustainable world or to greater prosperity. Yet, living a modern life devoid of any material possession or in the absence of any product, service, or amenity that has an impact on the environment feels improbable. Each day, we all require basic accoutrements, including food, clothing, and housing — and at some point, we all very likely also need access to transportation and medical care. Living with and practicing principles for planet pragmatism does not mean that we must live without or have less than. Planet pragmatism is not about pure survival, although that is an outcome. Rather, planet pragmatism provides us with wisdom, tools, and a playbook for making informed decisions in the moment, so that we can thrive with what we have, as we continue to pursue greater prosperity.

No one company, including the likes of Tesla, IKEA, Patagonia, Toms of Maine, Apple, Amazon, Google, Microsoft, GE, Wal-Mart, Starbucks, Nvidia, IBM, Unilever, Toyota, Nestle, or Nike, is 100% sustainable. Each of these multinational giants and thousands more are, however, pursuing or practicing principles of planet pragmatism that are slowly yielding a more intelligent industrial ecosystem and conscious consumer economy. To be clear, we need the ingenuity and innovation of sustainable enterprise, in all forms (i.e., small business, government agencies, non-governmental organizations (NGOs), multinational corporations (MNCs), state actors, public-private partners, and others) to collaborate on solving the world's toughest challenges including the restoration of nature, and provide equal opportunity for all people to pursue their definition of prosperity.

From a corporate perspective, Tesla has found a way to make a thriving business out of transportation electrification. Microsoft is dematerializing[42] and decarbonizing datacenters, striving to meet their net-zero goals. Patagonia is investing in regenerative organization practices that improve soil health and reduce greenhouse gas emissions. The company continues to move toward 100% renewable and recycled raw materials for their products. The company's goals[43] are admirable. By 2025, Patagonia will eliminate virgin petroleum material in their products and only use preferred materials; by 2025, their packaging will be 100% reusable, home compostable, renewable, or easily recyclable; and by 2040, Patagonia will be net zero across their entire business.

With earnings greater than $680 billion[44] (2025), the largest company in the world by revenue, Wal-Mart, has introduced consumers to more widespread environmental labelling and product disclosure. The retail behemoth is on a mission to eco-label[45] every product in its stores, pushing an eco-rating system and its requirements down to their suppliers. Further, Wal-Mart continues to provide useful resources to its suppliers, vendors, and customers such as its Recycling Playbook.[46] Wal-Mart's Recycling Playbook is a tool and useful guide for companies that seek to set up recyclable packaging and recycled content goals, and who are preparing for Extended Producer Responsibility regulations. Wal-Mart created their Sustainability Hub (see, https://www.walmartsustainabilityhub.com/) as a tool for their suppliers, connecting them with a plethora of resources that directly tie to principles of planet pragmatism such as resource moderation, temperance, and innovation.

Known for its ready-to-assemble furniture and household goods, the Netherlands-based IKEA has become a globally respected brand. IKEA is known for its modest-and-modern design, lower-cost quality products. The company's innovative ready-to-assemble products are akin to "LEGOs for adults," providing consumers with an easy-to-understand, step-by-step process to build their own furniture. The consumer assembly of products was and remains a business model innovation for IKEA. The company introduced an innovative packaging method called flat packs[47] which revolutionized the way many companies shipped their products directly to consumers. Flat pack shipping reduces packaging waste, optimizes shipping space, and results in lower energy consumption and greenhouse gas emissions. IKEA has progressively pursued sustainability focusing on three core priorities: (1) healthy and sustainable living; (2) climate, nature and circularity; and (3) fairness and equality. IKEA customers, many who are younger in age, value environmental conservation and protection. The company's leadership has acknowledged their intention to align the principles of the enterprise with those of their customers. In a 2019 interview, Jesper Brodin, chief executive of Ingka Group (the largest franchisee of IKEA stores), commented, "climate change and unsustainable consumption are among the biggest challenges we face in society.[48]" In the 1980s and early 1990s, the company was highly criticized for its use of formaldehyde in its products. Following public scrutiny and scandal, IKEA began a journey to embrace environmental stewardship. In 1990, the company adopted the Natural Step framework, which was founded by the Swedish cancer scientist, Karl-Henrik Robèrt[49]. Shortly thereafter, in 1992, IKEA developed and adopted its ten-point Environmental Action Plan, which catalyzed the company's next phase of progressive environmental action. As the world's largest buyer and retailer of wood, IKEA has been highly criticized for its timber management and forest stewardship efforts. The company owns about 136,000 acres of forest in the United States and approximately 450,000 acres in Europe. IKEA's 2021 Sustainability Report[50] claims that 99.5% of all wood that the company uses is either recycled or meets Forest Stewardship Council standards. Today, IKEA is working at the forefront of corporate supply chain sustainability, renewable energy development and use, product stewardship across the life cycle, responsible resource management and sustainable forestry, transportation electrification, waste minimization, recycling, and reuse. IKEA has an objective to make its entire supply chain climate positive by 2030.

Tesla, Patagonia, Wal-Mart, and IKEA are not 100% sustainable by any means. However, they are, through product, supply chain, and operational innovation, advancing the sustainability of their enterprise, and by virtue, their overarching impact on consumers and the planet. The sustainability efforts of these and hundreds of other MNCs should not be discounted. Humanity's path toward greater prosperity is entangled with capitalism and a deeply interconnected global supply chain and economy. Innovation influenced by and focused on sustainability has and will continue to play a significant role in pursuing prosperity through planet pragmatism.

The fact remains that a significant majority of the world's largest companies (and regional economies) are represented by resource-intensive (i.e., energy, water, minerals, natural resources) extractive industries, including oil and gas, chemical refining, mining, metal and materials processing, fishing, and global shipping. Although companies within these resource-intensive sectors have adopted best practices in environmental stewardship, natural resource management, energy efficiency, and sustainable production — these sectors are highly complex and require significant investments in infrastructure and materials processing, making them extremely difficult to decarbonize. Thus far, as our society has become more technologically advanced, our global demand for resource-intensive industrial production and commodities has increased. Despite reasonable corporate and governmental efforts to minimize pollution and environmental degradation, global industrial production continues to inflict a substantial adverse impact on public health and the natural environment. This does not mean that we should throw our arms up in defeat!

Beyond Business: Prosperity Through Planet Pragmatism is Ultimately in Your Hands

Corporate sustainable production is an essential component within the Planet Pragmatists' Playbook. Further, business is often the purveyor of creating value from innovation, that is, scaling the development and introduction of new products and technologies that solve problems, including sustainability related risks and challenges (i.e., dematerialization, clean energy and energy efficiency, circular material flows, clean water and air, eco restoration, infrastructure resilience, and so on). But we cannot (and shall not) rely solely on the merits of capitalism and free enterprise to solve humanity's sustainability crisis. Much of the past thirty years has been focused on addressing sustainability within our industrial base through global business. Working toward supply chain sustainability remains an enterprise objective, but this, like many other business sustainability challenges, is a self-induced problem born out of a global economy primarily operated by MNCs. While the world's citizens and consumers can benefit from a more sustainable supply chain, let's make no mistake, the real beneficiary is the corporation. Society absorbs the social and economic toll of unsustainable consumption, while nature (and humans) absorbs pollution and poison. There has been a narrow-minded focus to solve for sustainability through the lens of business, which has distorted our thinking on how to "solve for X" across many domains.

Attaining prosperity does not mean that we all live in mansions, drive exotic cars, vacation on super yachts, and adorn our bodies with luxury products. Excessive materialism is an outdated, false, and misleading image of prosperity. Planet prosperity, on the other hand, provides connection and belonging, meaning and purpose, faith and freedom. When we let go of pursuing and showing our wealth in material goods and embrace prosperity through peace, love, and understanding — the world becomes a much more welcoming and richer place. For nearly half a century, significant contemporary emphasis for improving the natural environment has been focused on the role of business. This was logical in the wake of the 1960s and 1970s environmental advocacy and policy movement that rose in response to environmental hazards and risks brought forth from industry. The question worth asking now is, does society place as much attention as citizens and consumers, in practicing planet pragmatism in our daily endeavors?

In response to negligent corporate behavior and pervasive environmental risks, we've put environmental laws and consumer protections in place. Companies and governments have worked to clean up contaminated sites and mitigate environmental risks. A new era of socially responsible enterprise has introduced environmentally benign products and embraced sustainable production of goods and services. But have these business-focused policies, regulatory, and incentive-based efforts on business been enough to transform the consumption-based, polluting, and wasteful global economy into one that can be protective of and restorative to nature?

The pragmatic potential and power in business ingenuity, entrepreneurship, and innovation are critical tools that solve problems and contribute to our shared prosperity. However, we must be cognizant of the challenges and limitations of free enterprise and the potential for placing false hope in our faith on business sustainability. To regain control of our prosperity, we must be willing to ask tough questions, make difficult decisions, and leverage our critical thinking, innate wisdom, and pragmatism in all that we are and do, well beyond the influence of business and free enterprise.

From a personal and pointed perspective, ask yourself, have I shown moderation, restraint, or temperance in my consumptive behaviors? Have I pursued alternatives to any products or services that do harm to public health or the environment? Am I living my values and principles in all that I am and do, not only as a consumer, but as a proactive global citizen willing to advocate for and practice principled prosperity and positive change? What can I do to simplify my life, yet enhance my health and well-being?

Points on Pragmatism

- *Change is omnipresent. The future and fate of business and society reside in our capacity to accept, adapt, learn, grow, and lead through change.*
- *Sustainability is not just a corporate scorecard relegated to quarterly earnings, reputational jousting, and securing new investments. If conceived and deployed as a foundation for adapting to change, enhancing quality of life through innovation, and enabling a new generation of principled leadership to emerge — then sustainability will finally and unequivocally "be in the moment, and have its moment" toward achieving a new prosperity for all.*
- *No singular corporation is sustainable. No industrial sector is sustainable. And the power and might of sustainable business will not exclusively save "people and planet." To regain control of our prosperity, we must be willing to ask tough questions, make difficult decisions, and leverage our critical thinking, innate wisdom, and pragmatism in all that we are and do, well beyond the influence of business and free enterprise.*

- *Sustainability happens in the moment. Our prosperity also happens in the moment. The choices we have in the present, as individuals and as a collective society, are the remnants of prior good and poor decisions that have been made. We have the power to manifest a brighter and more prosperous future by making more informed and wise decisions today. Pursuing an "in the moment" sustainability posture can and will guide a principled path toward a new prosperity.*

PART III

PLANET PRAGMATISM

7

THE PLANET PRAGMATISTS' PLAYBOOK

Adaptation: An Ancient Tool for Survival, A Modern Tool for Sustainability

Our ancestors invented and perfected the need to adapt. They did so for survival. Ancient people lived intimately with the planet. To shield themselves from the elements, they sought shelter. Eventually, they developed the skills and discovered how to construct their own shelter for protection. To stay warm, they wrapped themselves in earth-based materials, eventually giving rise to clothing. To nourish their bodies they foraged for food, quenched their thirst from the stream, and eventually hunted and then domesticated animals. Humans also discovered fire, which provided them with the utility of heat and the ability to see in the darkness of night.

Humans adapted to their surroundings, seeking at first, survival. But as humans' knowledge and skills for survival were practiced, they began to refine what they had created (shelter, heat, clothing, food) into shared utilities that also delivered greater comfort, joy, and opportunity for more people. We all know this story of human resilience well. We were taught this story in grade school, but we also lived this story. We ARE this story! Embedded within our DNA is a code for survival. The knowledge, skills, and capacity to learn, evolve, and innovate are embedded within us. And while our embedded sensors are always "on," recently, they have been signaling to our brain that it is time for our awareness to shift beyond alertness to act.

It's no secret that we are experiencing a greater frequency and amplitude of severe weather-related events and natural disasters. The onslaught of wildfires, atmospheric rivers that contribute to concentrated rain and snow events, floods, hurricanes and tornadoes, landslides, extreme heat and drought, and other climate risks are having an enormous impact on our infrastructure, communities, economy, and quality of life. In recent years much of the global political, policy, and corporate attention regarding climate risk has been focused on mitigating greenhouse gas emissions, particularly carbon dioxide.

Although the need for a near-term adaptation strategy has been elevated and discussed by scientists, policymakers, engineers, economists, and others concerned with climate risk, mitigation through carbon reductions, capture and storage, and displacement have taken center stage. It seems that while the warning signs of escalating climate risks have been touted, the urgency to act in this moment through adaptation strategies has been overshadowed by global interests that seek to curb carbon emissions in the future. I believe there is an element of human nature that interferes with the urgency for and strategies to support adaptation. Adaptation is focused on the need to act now, in the moment, to minimize, compensate for, or adapt to clear and present threats associated with climate risks. Mitigation, on the other hand, directs our attention to modifying society's behavior so that we reduce our production, or altogether eliminate greenhouse gases (GHG) from the atmosphere, to prevent the further proliferation of adverse climate risks to society. The image below summarizes the categorical distinction between adaptation and mitigation, providing a few examples. The image also illustrates how they are used as essential tools (gears) for planet pragmatism, reinforcing the network and interactive effects of all tools (solutions).

Five strategies that humanity can use right now to adapt to climate risk include: early warning systems, ecosystem restoration, climate-resilient infrastructure, water supply and security, and integrated long-term planning. These five climate adaptation strategies are endorsed by the United Nations Environment Programme[51] (UNEP). Further, these strategies align well with the "planet pragmatism" playbook. As shown in the graphic below, these five strategies, as well as opportunities for climate mitigation, align within the planet pragmatism "3P" plane that provides a Preventive, Predictive, and Proactive posture for evaluating and prioritizing the best-suited solutions for addressing climate risk. The "3P" plane is also a tool within the Planet Pragmatists' Playbook to support applied strategy and decision-making for pursuing prosperity through planet pragmatism.

Planet Pragmatism = Adaptation + Mitigation

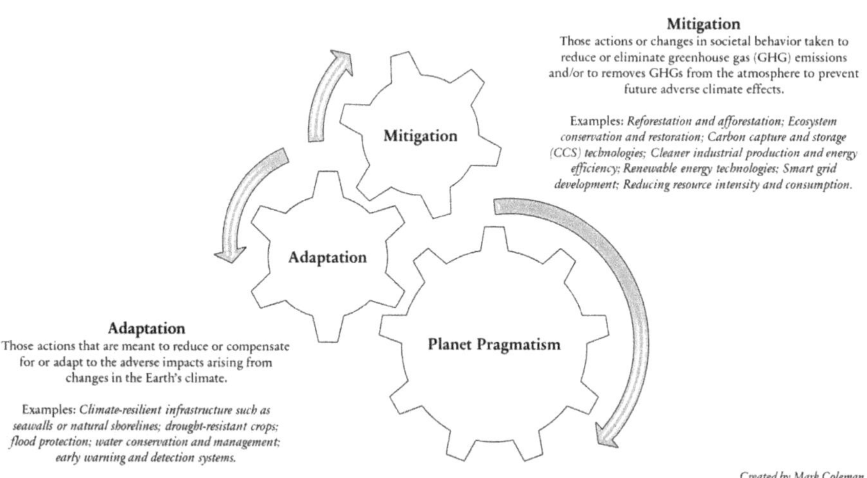

Mitigation
Those actions or changes in societal behavior taken to reduce or eliminate greenhouse gas (GHG) emissions and/or to removes GHGs from the atmosphere to prevent future adverse climate effects.
Examples: Reforestation and afforestation; Ecosystem conservation and restoration; Carbon capture and storage (CCS) technologies; Cleaner industrial production and energy efficiency; Renewable energy technologies; Smart grid development; Reducing resource intensity and consumption.
Mitigation
Adaptation
Planet Pragmatism
Adaptation
Those actions that are meant to reduce or compensate for or adapt to the adverse impacts arising from changes in the Earth's climate.
Examples: Climate-resilient infrastructure such as seawalls or natural shorelines; drought-resistant crops; flood protection; water conservation and management; early warning and detection systems.
Created by Mark Coleman

Five Ways to Adapt to Climate Risk Across the Planet Pragmatism 3D Plane: Preventive, Predictive, Proactive

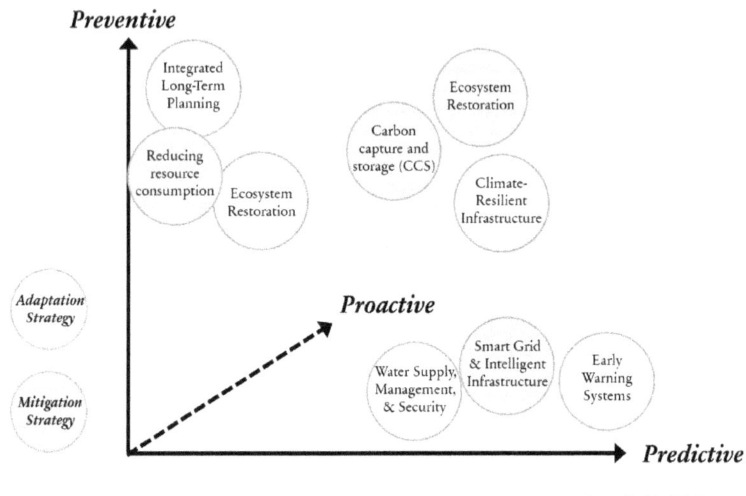

Created by Mark Coleman

For humanity to address its contribution and role in climate risk, we must focus expediently on both mitigation and adaptation with the same level of resolve. With destructive wildfire and high tide at our doorstep and poor air and water quality degrading our health, our innate fight-or-flight sense has kicked in. Communities are burning, children are having trouble breathing, food supplies are being disrupted, seawalls and levees are failing — the human-built infrastructure that enables comfort and quality of life to our modern version of survival is under assault. We now find ourselves once again, as we have previously experienced across the history of human civilization, being required to adapt or perish.

Although an individual human's sense of adaptation is born out of the necessity for survival, from a government and business perspective, adaptation can be a complex and multifaceted endeavor. Any organization or enterprise that creates value in the service of a customer is always susceptible to internal and external forces that influence how the organization optimizes resources to achieve that objective. Businesses that fail to adapt to shifting customer needs and preferences, for example, are vulnerable to competition and disruption. Government organizations and agencies that don't adequately or efficiently shift resources to the evolving needs of society fail to serve their constituents. Ultimately, that may result in a shift or change in power, either those that are elected to office, or those that were leading the governance and operations of the organization that failed to adapt. These are examples of how organizations (and their leaders) adapt to external influences. Business and government enterprises are also influenced by internal and external forces that require the organization to adapt. The physical risks associated with severe weather and climate shifts require organizations to think more holistically, strategically, and systematically about how they may be impacted by such risks. For example, if an organization owns and operates a portfolio of real estate which is critical to its operational success, logical questions come to mind to understand their readiness to respond to internal and external threats.

Where are their assets located geographically? Have weather events caused any operational disruption in the past, such as water intrusion or flooding, wind damage, sustained power outages, storm debris that restricted facility access, other scenarios that have previously impacted facility operations for employees, customers, suppliers, and vendors? What is the forecast for weather events for this location into the future? Have severe weather incidents been increasing? Has the enterprise seen any increase in facility or operational disruption or downtime? Has the facility had to make any repairs associated with physical weather or climate risks? Has the facility needed to make any new investments, such as in back-up power generation, flood mitigation, or other preparedness or post-event reconstruction services?

In March 2024, the MSCI Sustainability Institute put artificial intelligence (AI) chatbots to use, scouring publicly available corporate data and reports to decipher how frequently companies reported any activity on adaptation. The MSCI study[52] found that only 11% of globally publicly traded companies reported some activity on adaptation for their business. Given the heightened threat of climate risk and disruption to global enterprise, this statistic is alarming. When I think about enterprise sustainability, sure, what they are doing to mitigate GHGs is a top issue and concern, but so too is their ability to envelop a predictive, preventive, and proactive culture and operating system for responding to risks in the present moment, including climate and weather-induced. Enterprise stakeholders (i.e., investors, regulators, policy makers, board members, customers, vendors, employees) desire (and have come to expect) organizations that are robust and resilient, ready to respond to risk and change, and that can deliver a certainty of performance. In all reality, this is a lot to ask of any enterprise, but stakeholders are increasingly placing more value on those organizations that demonstrate they can perceive, identify, respond, or react and adapt to risk and change proactively and productively.

Since 2010, global investment in businesses that seek to decarbonize, reduce pollution, and generally become more sustainable has skyrocketed. Conversely, investors have not given as much attention to adaptation as a category of business and infrastructure investment in the face of climate risk, weather vulnerability, and uncertainty. The pace and scale of climate-based risks have been escalating. Wildfires occur naturally, and they serve a role in ecosystem health. However, studies[53] have shown that there is a direct correlation between our changing climate and the frequency, seasonal length, and total burned area of wildfires. Eco restoration is a practical wildfire adaptation and mitigation solution. The potential for eco restoration to deliver compounded interest to nature and humanity is explored in the sidebar below.

Eco restoration delivers compounded interest for nature and for humanity.

Eco restoration and nature-based solutions yield returns for the resilience of the planet and our communities. Our world is increasingly feeling inhospitable. It is difficult to conceive and pursue greater prosperity when communities are washed away by floods, burned to ashes, or fail to provide the basic requirements for clean water, affordable and clean energy, and food security. In recent years, the focus on climate change and the reduction of carbon emissions has taken center stage as a global imperative.

Decarbonizing our modern economy should be an essential strategy for our planet pragmatism playbook. But decarbonization of the global economy is taking considerable time, financial capital, and natural resources. While many people, myself included, would like to move fast, the reality is that the global economy will, in the absence of major breakthroughs in the generation of clean and affordable energy, transition over the course of several decades. This said, the transition to a clean energy economy is on a concerted, rapid global ascent, bringing forth new technology and innovation that delivers reduced carbon or no carbon energy to fuel and power our modern society.

According to data and projections from the International Energy Agency (IEA)[54] and other sources, it will take time for the global economy to fully decouple from carbon-intensive fossil fuels. In fact, the Organization of Petroleum Exporting Countries (OPEC) estimates[55] "global primary energy demand will increase 24% by 2050, driven by the non-OECD. Global primary energy demand is set to increase from 301 million barrels of oil equivalent a day (mboe/d) in 2023 to 374 mboe/d in 2050, an increase of 24% over the outlook period."

Meanwhile, atmospheric concentrations of carbon dioxide and other greenhouse gases (GHGs) continue to increase due to the time it takes for them to naturally break down into their molecular components. Subsequently, the frequency, scale, and severity of climate-related risks continue to amplify and wreak havoc on our infrastructure, communities, and pursuit of prosperity. While our climate risks can feel overwhelming, there are tried and tested solutions at our disposal that can mitigate climate impacts and heal the planet's ecosystems simultaneously. Eco restoration is a process of human intervention to "assist in the recovery of ecosystems that have been degraded or destroyed, as well as conserving the ecosystems that are still intact.[56]"

The process of eco restoration enables natural solutions for carbon mitigation while providing adaptive and resilient ecosystem services. Examples of eco restoration include: reforestation (planting trees in deforested areas), removing invasive species, reintroduction of native plants and animals, wetland restoration, coral reef rehabilitation, stream and prairie lands restoration, and the use of green spaces in urban areas. Eco restoration supports the regeneration of the planet's natural ecosystems, which deliver life-essential ecosystem services such as provisioning for food and clean water, clean air, flood and drought mitigation, natural resource commodities including timber and fisheries, among many more.

Ashes to Ashes: Do We Have What It Takes to Get Climate Adaptation Right?

With ashes on the soles of our feet and deep sorrow on the souls of our minds, we need to take this moment of despair and learn from the LA wildfires. The planet, our home, is under assault and changing.

We need to restructure how we live, work, and play in a manner that is adaptive, resilient, and sustainable. This will challenge every notion of our consumer culture and how we've previously built infrastructure in America.

We have available to us today, the necessary technology, tools, and intellect to evoke a predictive, preventive, and proactive posture on how communities can be rebuilt to foster a new generation of American prosperity and ingenuity.

Can we evoke dignified leadership, collective action, and common sense for the common good to rebuild and revitalize the LA region, and build a stronger America?

Sustainability Upended: Climate Mitigation was Never the Full Story

The most recent and devastating Los Angeles region wildfires[57] should challenge everyone's notion of sustainability. With our rapidly changing climate, we are woefully failing to keep pace with the renewal, resilience, and adaptation of infrastructure that is necessary to keep us safe and secure. Underlying our deficiencies for infrastructure sustainability is the over-politicized bravado, egoism, and gotcha moments, debating whose right or wrong, characterized by our pervasive political culture.

It's sadly clear that even in moments of great need, the divisive rhetoric remains constant, pilfering our attention on what matters most. Meanwhile, millions of people remain starved for dignified leadership that can lead during these uncertain times. In our debate-first driven society, our infrastructure is not failing us, we are failing ourselves. Too much time and effort are spent pointing fingers, placing blame, and creating distrust as opposed to solving our real sustainability (social, economic, and environmental) challenges.

The LA wildfires are evidence that our sustainability is not improving, rather, it is rapidly declining. Although society has had its fair share of unfathomable natural and human-provoked disasters, we have failed to evolve. The LA wildfires represent another clear indication that to survive, let alone thrive in this rapidly changing world, we must change. More than ever, we need pragmatic, common sense, logical leaders who can guide us out of the flames toward a new shared prosperity.

The Past is Behind Us, the Future is Defined by Our Decisions Today

Perennial optimists, Americans in particular, tend to believe that most issues will go away or resolve themselves over time. For decades, Americans believed and lived as if they could have their cake and eat it too, particularly when it comes to consumption, economic development, and regional growth. Yet, here we are choking on the cake, wide-eyed and fearful in the face of imminent destruction and death. Time and time again, event after horrific event, we watch, we debate, we shrug it off as something that is isolated and will pass us by. But we all know, cancer doesn't just pass by. You cannot simply pray and your cancer will go away. You need to reconcile your fear with the reality of change and act.

If left untreated or ignored, cancer proliferates from within the body until it spreads, like wildfire, and destroys any hope of recovery. For far too long, America has had "consumption cancer," that is, a belief that we can buy and build our way out of anything and then behave accordingly. This cancer can be characterized by many nonsensical behaviors and epitomized by many material things — look at our car, house, and food culture for plenty of examples. There is no point in shaming, blaming, or berating specific generations, individuals, or institutions at this point in our diagnosis. What we need to do is take this moment to take stock of our health, acknowledge our shared condition, and redefine our current reality and future prosperity accordingly.

Achieving this will require tough decisions and, perhaps, an entirely new way of how we conceive modern life, including what is most important to us as individuals and as a collective society. It also beckons us to reimagine consumption, most likely by moving beyond what's trending. We can't bury our heads in the sand and pretend our consumption culture is not overconsuming us.

The signs were all around us. The LA wildfires are not AI-generated. This is the real deal, and there was no drill. The shocking scale of impact of the LA wildfires was sparked and fueled by the confluence of numerous factors: high winds, prolonged drought, densification and encroachment of housing near wildlands, outdated water infrastructure, and the generalized human element of ignorance and skepticism. Although all the environmental conditions were right for a massive wildfire, likely very few ever thought, "This would happen to us."

Our generation is on the front lines of a planet and society under duress. Our world is increasingly facing serious threats promulgated from climate, energy, financial, security, technological, and other risks. Many of these risks are human-made and induced. Others are truly acts of God. In either case, the prosperity, freedom, and fate of society lie in the balance. We must recognize that the years of eating cake are over. We need to start making tough decisions regarding the trade-offs that are before us.

Refocusing our Resources on the Problem at Hand

In responding to news coverage of the LA wildfires, my 16-year-old son asked, "Why isn't there more water available from the municipal pipes? Why can't the firemen just use ocean water? Is there a way to remove salt from ocean water?" His questions turned into a thoughtful thirty-minute conversation between him, Mom, and me. My son's line of reasoning was not wrong. He was asking the right questions, seeking solutions. Our family conversation, backlit by the background news testimonials from people on the frontlines, including the heroism of first responders and the immense sense of loss felt by those who lost their homes, turned to what happens from here. "Why can't we build a huge desalination plant in LA, like they have in Dubai?" asked my son.

Our family conversation reawakened my mind's eye toward a topic that has recaptured my thought and attention on the elusive allure of sustainability in the past couple of years. Essentially, I view sustainability as an all-encompassing pathway toward greater prosperity. Sustainability is more than "solving for X," like climate risk through decarbonization. We represent a sustainability generation in transition, one foot in the past, one in the present, and a reach to a future that can feel uncertain. The future is, as it always has been, what we make of it. Our intentions and how we act in the here and now determine what comes next. Thus, sustainability is not represented as a singular goal or objective, but as an underlying construct of transition and transformation. It is the epitome of change management.

In that, there are trade-offs and compromises that must be evaluated and prioritized. I don't believe that a singular technology fix or government mandate will "save us." I do believe however, that we have the capacity and intellect to improve the world through innovation, in all its forms. There is no limit to human ingenuity. While society seems to be leaning heavily on the potential of artificial intelligence (AI), underlying the optimism tied to the automation of everything is an innate fear that humans will be rendered less useful, less intelligent. Yet as we sit here today, we represent the intellect that is fashioning the very fate that we fear.

Let's assume for a moment that the LA wildfires are not an anomaly. The frequency, scale, and severity of climate disasters have been steadily increasing for some time now. There is no need to assume here, we know this. We've lived this. Let's also factor in that with 8+ billion people on the planet, competition for resources has intensified, particularly over the past three decades, and it will continue to escalate in coming years. Again, this is well understood.

So, our collective home is under great duress. In recent years, significant investments in new technology, capital operating expenditures, and development of policy mandates have been placed on reducing the emission of carbon dioxide (CO2) and other greenhouse gases (GHG) into the atmosphere. Undoubtedly, our modern society, and in particular the developed nations that emit the majority of GHGs, need to curtail the inefficient combustion of fossil fuels and curtail and mitigate carbon and associated GHGs. But although decarbonization seems like it should be simple, it is riddled with needless complexity and contradiction.

Time to Assess the Sustainability Trade-Offs of Where and How We Invest

With the LA wildfires on full display, the paradox we are living in can be seen with greater contrast. We need to decarbonize, yet that takes time, and at this point, the impact of reducing carbon today will not be felt by many regions and people for decades. And then there are thousands of people in LA, now homeless and categorically defined as climate migrants, who need relief and a solution right now. The necessity to invest in climate adaptation has never been greater. However, our resources, finances, policy, and leadership intentions have been ill-aligned with the immediacy of the fire that is at our doorstep.

If I did the research to scour the soundbites of each major climate disaster over the past thirty years, I'm very confident that for each one, there is someone who has previously said, "This should be a wake-up call for society." But here we are, eyes wide open, yet still not really awake. It is time to rethink our allocation of resources and leadership intentions to better align with the real challenges people and communities are facing now, and have been facing, for decades. We've kicked the proverbial can down the road long enough. In fact, we've run out of road, and there is nowhere left to go. We find ourselves pegged between a vast ocean and steep mountain, walking amid the ash, questioning how and why.

Pan out from the LA wildfires, and there remains a steady stream of business news celebrating the astronomical financial performance of AI companies over the past year. Business analysts predict continued growth in tech, led by continued innovation in AI. But they caution that the Achilles heel or limiting factor for continued exponential growth is directly tied to the availability of reliable, affordable, and clean electricity to power datacenters and advanced manufacturing facilities.

The State of California has had significant electric power utility challenges ranging from power quality and reliability, ratepayer affordability, lack of clean generation assets, power shut-offs and rolling Blackouts. California's energy grid is strained and overtaxed by high demand, climate risks including high winds and wildfires, and aging infrastructure, among other concerns. California's ongoing energy crisis[58] and water crisis[59] are not mutually exclusive. They are intimately intertwined with severe weather and wildfires. The need for adaptive, resilient, and sustainable infrastructure solutions in California has never been greater.

The promise of advanced technology, including AI, is exciting for investors and tech enthusiasts. But when you don't have access to basic energy or water services, or the ability to be properly housed, the allure of AI does not shine as brightly. I reference this not as an assault on AI or advanced tech, but as an observation on where society and investors have and continue to place their attention. We have enormous potential, here and now, with cost-effective, proven, and pragmatic technology and solutions, to design — build — maintain — and sustain an infrastructure that can be adaptive, resilient, and sustainable. Getting there, will require a wholesale change in mindset, on where we place leadership intention, policy and financial incentive, and community building.

Our society has traditionally built and viewed major public utilities as discrete value streams, for example, energy/electricity, transportation/mobility, telecommunications, education, security, and so on. Over the past century, advancements in technology and behavioral changes in how society engages with public utilities have not kept pace with each other. Today, we have the potential to converge the value streams (i.e., the resources, operations, benefits, and impacts) of previously siloed public utilities in ways that can dramatically reduce costs, enhance efficiency, and revolutionize how the infrastructure that serves us can become adaptive, resilient, and sustainable. We have an opportunity to integrate advanced technology, including AI and clean energy technologies, with innovations in mobility, manufacturing, and housing to create restorative, adaptive, resilient communities.

Human Dignity Rises from the Ashes of Despair

The impact of the LA wildfires was nondiscriminatory. People from all socioeconomic backgrounds lost homes and personal possessions, and in some tragic cases, friends and loved ones. As the California wildfires continued to rage, the economic impact on the region continued to rise to an estimated $100 billion in economic damage. For the people dealing firsthand with wildfires, the social and psychological toll far exceeded any immediate financial loss.

Much like other mega-disasters of our time, the LA wildfires have, almost immediately, catalyzed a global response. The courage of frontline first responders and connected stories of desperation and loss have spread a wildfire of compassion across the country. Young adults have launched funding campaigns. People across the LA region have banded together to provide food, clothing, and shelter. Business leaders have stepped up to provide resources and services to aid in the immediate needs and the longer-term recovery of the people, and a region, in shock. Time and again, human dignity rises out of the ashes of despair and suffering. At our core, people are good. People are resilient. People succeed and thrive when the community is strong. *Can we evoke dignified leadership, collective action, and common sense for the common good to rebuild and revitalize the LA region, and build a stronger America?*

Points on Pragmatism

- *Our modern and consumer-focused society has pitted humans against the laws of nature. Although ancient wisdom for living within the bounds of nature is encoded in all of us, that knowledge has not transcended through all that we are or do. Too often, we view ourselves as separate from, or above, the beautiful and intricate foundation of nature, which enables us to have a place in the cycle of life. Ancient people understood and valued life universally. They saw themselves integrated and integral to nature and vice versa. When we finally accept and stop fighting against this truth, humanity will evolve in spiritual and metaphysical dimensions that enable our future prosperity and enlightenment.*

- *Adaptation is an ancient tool for survival and a modern tool for sustainability. Adaptation and resilience represent examples of planet pragmatism. Grounded in a need for survival, these ancient tools provide us with the knowledge and agency to identify risk, navigate change, and problem-solve through innovation.*
- *Human intuition is intrinsically connected to nature and with the Universe. We must learn to tap into our senses, evoke the knowledge encoded in our DNA and ancient wisdom, and listen more intently to the signals that are continuously shaping our world and our reality. When we engage with and embrace nature, we get closer to understanding ourselves and where our future path will lead.*

8

COMMON SENSE FOR THE COMMON GOOD: CRACKING THE CODE ON CLIMATE CATASTROPHE MAY BE SIMPLER [AND CLOSER] THAN WE THINK

Have you ever noticed that humans tend to overcomplicate life's simple things and attempt to oversimplify the complex? Our interpersonal and societal relationships and behaviors are full of contradictions. War and peace, addiction and recovery, love and hate, success and failure, disease and cure, marriage and divorce, birth and death — our lives fluctuate between polar extremes that continuously push us together and pull us apart. It's no wonder that when it comes to all facets of sustainability, we are a hot mess.

How we frame, discuss, and proactively pursue tough sustainability challenges like climate risk, energy equity, and severe declines in global biodiversity illuminates how we flow between the polar extremes of overcomplication versus simplification.

For example, in the energy efficiency space, there is the adage, "a kilowatt-hour saved is a kilowatt-hour earned." We can, without sophisticated technologies, processes, or controls, have an immediate and direct impact on reducing energy consumption, pollution, and associated environmental externalities. Of course, with our vast intellect, we can also overdesign, engineer, and deliver complex technologies and create even more complex programs to accomplish the same task. Increasingly, our lives are being disrupted by climate and ecological risks.

Common sense, defined as "good sense and sound judgment in practical matters," is a characteristic that we should all have, by now, yet seldom do we exercise it with as much enthusiasm as ego or hubris. For humans, I would argue that there is nothing more practical than survival. One would think that over the compounded acquired and lived time humans have had on Earth, we must have some common sense. We do, but it does not seem to permeate into all that we are or do. Humans have a critical flaw in that we are tireless tinkerers. We do not like to leave well enough alone. In our pursuit of [more] knowledge and understanding and greater prosperity, we often forget age-old virtues such as moderation, temperance, and the practical pursuit of excellence.

I believe this is why much of our strife on climate change remains unresolved. We believe, perhaps more so than the simple elegance of a practical measure like consumer restraint, that we can engineer our way out of climate and ecological calamity. There is truth and precedent in our inventiveness, giving lift to our pomposity. Humans have devised some pretty impressive technological wonders in our time. Bridges, tunnels, skyscrapers, the Internet, medical diagnostics, space travel, and quantum computing — our ingenuity is quite impressive. In our modern culture, however, we seem to take greater pride in our engineering prowess and marvels than from the humility that could be gained by learning from our past failures. This attitude only fuels our arrogance.

The time has come for us to take a step back on ambiguous sustainability (including climate) targets and the tendency to overcomplicate sustainability solutions. The sustainability dilemma before us is not based entirely on the availability or scarcity of money, technology, knowledge, or know-how. We have yet to crack the code on sustainability and the looming climate catastrophe, largely because we have not yet coalesced the generational will and harnessed the full potential for acting with common sense for the common good.

To effectively address the sustainability risk of our time, we must embrace an ethos of common sense as an essential and core value of critical thinking, as well as one that is integrated into the foundation of how we pursue design, engineering, and technological advancement. In doing so, we just may have a shot at evolving in step with the planet's swift climate and ecological changes that are shaping our future.

Is a climate and ecological catastrophe at our doorstep?

Look no further than this year's hurricane season and the devastating impacts caused by Hurricanes Milton and Helene as evidence of increasingly severe and larger storms. The massive rainfall and inland flooding caused by Helene took many communities by surprise, particularly those in hard-hit regions of North Carolina. The rapid ascent and scale of Milton brought John Morales, a veteran Florida meteorologist, to tears[60], as he reported live on the growth of the behemoth storm.

On October 9, the World Wildlife Fund (WWF)[61] issued their Living Planet Report 2024, warning that parts of the planet have approached dangerous tipping points, measured by dramatic losses of biodiversity. The headline by WWF[62] stated, *"Catastrophic 73% decline in the average size of global wildlife populations in just 50 years shows a 'system in peril."*

According to the 2024 WWF Living Planet report, human activity associated with overharvesting natural resources and food systems has exacerbated habitat loss and ecological degradation. The 2024 WWF report further identified the spread of invasive species, disease, and climate change as critical convergent factors that reinforce a negative feedback loop further escalating ecosystem decline and the planet's biodiversity loss.

The most severe biodiversity decline, as measured by the Living Planet Index (administered by the Zoological Society of London), was measured in freshwater populations (85% decline), followed by terrestrial (69%), and marine (56%). The capacity for nature to provide the resources and services necessary for a healthy and vibrant planet, and subsequently, the survival of humans, is under assault. The combination of human and natural influences is clearly taking a toll on environmental quality and the capacity of the natural world to deliver the ecosystem services that have traditionally enabled human survival, i.e., the provisioning, regulating, cultural, and supporting services of nature. Additional information on ecosystem services is available from many sources, including the National Wildlife Federation[63].

Back to the question, *"Is a climate and ecological catastrophe at our doorstep?"* The answer, by most commonsense accounts, is a resounding yes. A climate and ecosystem catastrophe is most definitely at our doorstep. In fact, we can probably agree that it has made its way well through the front door, kept its muddy boots on, trampled through the house, sat squarely down on the cozy, clean sofa, kicked up its feet on the leather ottoman, and made itself right at home.

We have all played a supporting role in letting the intruder into our home

Climate and ecological risk and the potential for ongoing catastrophes are here to stay. An intruder is inside our home, wreaking havoc and making a mockery of us. However, unlike unwelcome household pests, our current climate and ecological risks cannot be exterminated by calling in a service technician. This is a job we need to figure out and do ourselves.

Right now, we are failing, miserably. The data, the persistent risk events, the economic costs, and the tragic loss of human life all indicate that we do not have control of "our house." As a metaphor, the "house" is our planet, but it is also the infrastructure and chosen paths by which we have designed and constructed to occupy the planet. In general, our modern civilization has attempted to cultivate land, tame nature, and make the planet our home. We've done so reasonably well and successfully. But the proverbial gloves have come off, and nature and planetary systems are fighting back with a vigor and ferocity that we have not previously encountered. We certainly have not thoughtfully designed and engineered our infrastructure for the intruder in our home and the ensuing fight that has only just begun.

Like any war, one must question the underlying logic of what we are fighting for and why the fight must ensue. Right now, we are caught up in and distracted by an ideological and political battle regarding climate risk and the plant's state of health. Meanwhile, the forces that are underway to shape the planet and our collective futures do not care about our cultural perspective and interpersonal infighting. The hypersensitive and politicized debate that continues to exhaustingly play out over who is right and who is wrong on climate change is an irrelevant sideshow to the enormous change happening right in front of us, right under our feet, and right within our homes. The classic idiom, *we're simply rearranging the deck chairs on a sinking ship,* comes to the top of mind.

We are waging the wrong war, fighting the futile fight, which is against ourselves. In this lose-lose battle, we have yet to reach a common ground on how we can remove the unnecessary obfuscation surrounding climate change and focus on commonsense actions that can bridge the chasm between our perceptions and the realities of a dynamic planet.

Utopia [re]found, *how lucky are [were] we?*

Our home has always been in a state of flux. Humans never needed to look for a utopia. It has, at least by the measure of human existence, always existed beneath our feet. **How lucky are we?**

How humans have taken advantage of this utopian environment

has, however, created a host of social, economic, environmental, public health, and humanitarian crises. Essentially, we are exploring the Universe and the potential to colonize other planets, including Mars, because we sense that what was once a pristine paradise may not continue to produce the fruits and freedom we have selfishly enjoyed.

This said, our occupation of Earth has been afforded by its own "Goldilocks" period of geologic, chemical, biological, and planetary ripening. And, oh boy! Did we arrive at harvest time or what?! To put it another way, humans basically won the Universe's "Mega Jackpot, trifecta box, and roulette wheel" all at once. All the necessary conditions to support our evolution and survival have been put in a mixer and blended just right for the past few million years. A few million years may seem like a long time to create paradise, yet it's but a fraction of the time that the planet has been busy building the right foundation to stand up for our home.

It has long been time for humanity to stop gambling away our future. The signs could not be clearer. For us to maintain (let alone improve upon) our home, we need to acknowledge and fully accept that climate and ecosystem risks have rooted themselves deep within our collective foundation. The climate and ecological challenges we face do not differentiate across political parties, religious affiliation, economic status, gender identity, or any other socioeconomic-cultural delineation. The planet has been our common ground. Now we must work together for the common good in the face of an uncommon adversary.

Today, here and now, we have the underlying wisdom, intellect, technology, and tools to adapt to a changing climate as we adopt a more resilient, sustainable, and restorative infrastructure. Guided by common sense principles for adopting a preventive, predictive, and proactive posture on how we pursue prosperity, we can create positive and pragmatic solutions that enhance our quality of life, deliver safety and security, protect and heal our (planetary) home, as we seek to ensure the resilience and sustainability of our personal home.

Therein lies the challenge. Human ingenuity is a powerful thing. Our rapid technological advance is giving rise to enormous power and new capability. Just consider the amazing potential of artificial intelligence (AI) and quantum computing. How we wield and put this power and capability to use (i.e., for the common good or as a division line among people) in the coming years will be a key determinant of our future prosperity.

The planet is asking us to evolve, *are we listening?*

We must shift our logic and reasoning away from questions like, "Are we facing a climate and ecological catastrophe?" and instead focus on cracking the code on common-sense solutions to advance today. We must pursue both temperance and restraint regarding overconsumption of resources, and we must equally pursue aggressive integration of knowledge so that we can transition our infrastructure and our homes to be more adaptive, resilient, and restorative.

There is currently no silver bullet technology that will change our trajectory to save us or the planet. Our ability to survive and thrive is predicated on whether we can live, learn, and lead together. This task, albeit the ultimate example of common sense, has always been, and remains, the greatest challenge of humans. We have so much in common, yet we create our own confusion and chaos. We have shown that when we work together toward a common goal, we can accomplish incredible feats, including cracking the code on disease, exploring the far reaches of the Universe, and designing more resilient and adaptive cities. So why then do we continue to remain stuck in a cyclical loop and culture of cynicism, distrust, and divisiveness?

The planet is evolving, and changes are underway at the hands of humans. This evolution and change do not (as a point of opinion) offer any less of a utopia for humans to embrace than it did yesterday, a hundred, a thousand, or ten thousand years ago. **Just as it has always been and shall remain, the planet and its capacity to provide us with prosperity come down to what we make of it.** What we make of it requires that we evoke common sense for the common good. We have the option to make wise, practical, and pragmatic decisions in the present, as opposed to creating unnecessary challenges.

The planet, our home, is beckoning us to evolve, emotionally, intelligently, and perhaps spiritually. Will we answer the call? Can we put aside our differences and find the courage to work collaboratively, take care of each other, and pursue prosperity, with common sense, pragmatism, and resolve?

Points on Pragmatism

- *Today we are too caught up in and distracted by an ideological and political battle regarding climate risk and the plant's state of health. A hypersensitive and politicized debate continues to play out over who is right and who is wrong on climate change is an irrelevant sideshow to the enormous change happening right in front of us, right under our feet, and right within our home. "Are we simply rearranging the deck chairs on a sinking ship?"*
- *There is currently no silver bullet technology that will change our trajectory to save us or the planet. Our ability to survive and thrive is predicated on whether we can live, learn, and lead together. This task, albeit the ultimate example of common sense, has always been, and remains, the greatest challenge of humans.*
- *Human ingenuity is powerful. Our rapid technological advance is giving rise to enormous power and new capability. How we wield and put this power and capability to use (i.e., for the common good or as a division line among people) in coming years will be a key determinant of our future prosperity.*
- *Just as it has always been and shall remain, the planet and its capacity to provide us with prosperity, comes down to what we make of it. What we make of it requires that we evoke common sense for the common good. We have the option to make wise, practical, and pragmatic decisions in the present, as opposed to creating unnecessary challenges.*

9

PLANET PRAGMATISM: OUR ROADMAP TO PRINCIPLED PROSPERITY

You can call this chapter a deliberate "timeout" from your daily regimen within the matrix. This is an opportunity to pause and reflect on the past, redefine what's needed and what's possible, and redirect our energy and purpose toward a new North Star, that is, future direction. Within the subtitle of this book is the phrase, "the new path to prosperity." I deliberately chose these words to be associated with pragmatism for this book to shift the pendulum of popular sustainability semantics back toward a place of common sense. Let me explain.

The title of this book, and for this chapter, incorporates three interlocking themes concerning the future of the planet and fate of humanity:

- **Pragmatism** — the innate capacity for humans to use logic and think critically while enveloping rational and realistic goals, expectations, and behaviors to address their needs in any moment.

- **Playbook** — the necessity for having a plan of action, that is disciplined yet adaptive, and incorporates the dimensions of change, time, resilience, and human ingenuity.
- **Principled Prosperity** — the intentional leadership construct and decision to design, build, and live a moral, just, equitable future — one which humanity's humility for life, the planet, the Universe, and the unknown leads to a continued growth of our potential and deeper understanding of our shared purpose.

These three themes are not intended to oversimplify humanity's relationship with the planet or as a construct to define the intricacies of human-to-human, human-to-planet, or human-to-universe interconnectedness. These themes are, however, interwoven to help us reimagine and recast our current thinking and future direction.

I've always taken some solace in knowing that we can, as individuals and as a society, change our story and trajectory. We can shift attention, attitudes, and behaviors. If we're not happy with who we are today, we can change. Humans are adaptive — sometimes we choose to be, and sometimes we are forced to be. Encoded in our DNA are the instructions for us to carry out life skills like observation, perception, intuition, fight or flight, and other behaviors or responses to our natural and human-built world. We've constructed a human experience that is enthralled with stimulating our five senses, putting them to use for pleasure as much as they were used thousands of years ago for survival. We've evolved and adapted. But in doing so, we've put ourselves in a precarious place. We haven't mastered survival, but in the hubris of our technocratic and more comfortable detachment from humans' early days of "fight or flight," we believe we can design our way out of anything. Our arrogance is our Achilles heel, and the planet and Universe are pulling quiver after quiver, taking aim at our survival.

First, let me begin by acknowledging a bias that cuts across my intellect, worldview, and perception of humanity. While I'd love for every human to interlock our hands and hearts and sing "Kumbaya my Lord" as we navigate toward a shiny North Star vision for the future, grounded by shared values and actions taken today, I don't realistically see that happening anytime soon. Ouch! That hurts to write, say, and admit. But, in my defense, corralling the individual beliefs, traits, cultures, and values of over 8 billion people is an overwhelming goal. This said, for several years, I've been working with some incredibly visionary and gifted people who are working to transform society on an enormous scale. Although this book approaches planetary boundaries and limits through a human-defined lens of pragmatism, that does not mean that we cannot drive transformational change at scale. What this book will reveal is that applying principles for pragmatism across multiple human-planetary dimensions (i.e., science, policy, technology, economics, finance, energy, transportation, etc.) allows us to engage, learn, grow, lead, and evolve with less risk, greater certainty of outcome, and with attainable results on our quality of life and that of planetary systems.

Secondly, let me state that pessimism is the enemy of the pursuit of better, that is, pessimism attacks faith, extinguishes the fire of creativity and invention, and drains courage from those willing to dare to challenge convention and lead toward higher virtues guided by shared principles. Observing the woes of the world, let alone reconciling one's own contribution or stake in manifesting and maintaining those challenges, if not atrocities, can leave one free-falling into the abyss of pessimism. If you've ever streamed any Breaking News for more than 24 hours, you know what I'm speaking of. Global society can appear to be broken. Wars and wildfires, demagogue politicians fuel and fan the flame of distrust, civil unrest leading to human migration and conflict, feelings of loneliness and depression driven by the pandemic, and recurrent global issues among our youth, and so many other social, economic, and environmental challenges indirectly and directly impact our lives. Whether we are on the frontlines or perched safely behind a screen receiving real-time feeds, these issues appear to be promulgating and persevering despite our greater awareness. As we are constantly bombarded by extreme external events, we feel helpless, afraid, and alone. Given the onslaught of challenges we consume through our screens each day, it's easy to throw one's hands up and want to retreat from society. Pessimism can creep into our minds and hearts and quickly give rise to outright panic and protest if we don't think critically and act pragmatically.

Humans are incredibly observant, keen to collect data on almost anything and everything, characterizing our behavior and our world in ways that we've never done before. Our interest in and thirst for data has given rise to artificial intelligence (AI). But without a North Star to guide the principled implementation of this powerful technology, humanity is merely gaming our present and gambling our future.

The morality of humanity occurs in the moment, that is, how we act in any situation is predicated on our values, perceptions, biases, and other factors. Our morality is also designed into the physical and digital infrastructure, products, and services we use daily. The decisions (i.e., design, engineering, economic, policy, development, etc.) we make today, including those that incorporate advanced technologies like AI, require us to regularly calibrate our moral compass. Morality is lived in the moment, but it is also a construct that cuts across time and even generations of people and lived experiences. History shows us that humans have flaws, and that we don't always get it right when it comes down to the implementation of technology and how it can enhance our quality of life and uplift society in an equitable and ethical manner. Incorporating an intentional "ethical foresight for the future" can help alleviate unintended consequences of our decisions today. Doing so will require us to have a preventive, predictive, and proactive mindset in how we can pragmatically infuse our values and ethics into our lives for beneficial use today, and with due consideration on what the longer-term impact may be for people and the planet in the future.

AI: Technological Hubris or Humble Servant for the People?

Generative AI has exploded onto the global scene in the past two years. As a tool that can minimize or eliminate mundane tasks,

Generative AI stands to deliver humanity an opportunity for optimizing our time and potentially enhancing our quality of life. Generative AI is but one category of artificial intelligence. Since AI has become popularized, many questions have arisen regarding the impact of this powerful technology. A few key questions include:

- Will AI lead to mass layoffs and loss of jobs?
- Will AI achieve a sentient state, on par with human consciousness? And if so, what would that mean for the future and fate of humanity?
- How might AI be used for nefarious and criminal activities?
- Can and will AI perpetuate pre-existing biases, prejudices, and/or inequalities among humans?
- Can we encode ethics and morality into AI in a ubiquitous manner that supersedes our own limitations and judgment on ourselves and each other?
- If we completely relegate our decisions to AI, robots, or other autonomous tools and technologies, are we essentially diminishing our morality?
- Will AI lead to the downfall of humans? Will popularized notions of AI taking over humans and the planet, much like as seen in movies including *The Terminator, The Creator, Transcendence, iRobot,* and *A.I. Rising,* result in our own manifestation of AI doomsday scenarios, many of which are already encoded within our subconscious?

In 2019, I travelled to Redmond, Washington, for a visit with Microsoft. It was a great meeting, a privilege really, to tour part of their innovation center and to discuss the future of technology. During a meeting with some senior leadership and subject matter experts, the question around AI and inherent design bias was raised. As an overgeneralization, many top talent AI engineers have a similar pedigree. They tend to be men, early to mid-twenties, who attended elite academic programs at top universities. In fact, over seventy percent[64] of the AI engineers employed in 2024 were men. Clearly, gender bias poses an enormous challenge given the huge potential AI poses in shaping the future of humanity. Gender bias, while an already stark reality, is but only one of a multitude of biases that plague the ethical and moral engineering and deployment of AI. If Microsoft and other technology companies are hiring very smart and capable software engineers from top academic institutions who are male, and who may even exhibit other shared or common characteristics (i.e., race, culture, religion, age, sexual orientation, economic status, and others), the probability of creating AI that has inherent bias built in is very high. When our design engineers, particularly software engineers, are not representative of the population (and customers) they are designing and building for, something is lost in translation. Although tech companies may not intentionally be building bias into their code, there is a high likelihood that unintended outcomes occur. Tech companies, including Microsoft, Google, IBM, NVIDIA, Amazon, and others, have taken steps to have greater diversity in their hiring. They have also attempted to have ethical and moral reasoning as a foundation for the AI they are developing.

The AI ethics landscape is evolving. Primary ethical concerns on the proliferation of AI include:

- **Bias and Discrimination** — ensuring that AI does not intentionally or unintentionally introduce bias or discrimination into its design or use.
- **Transparency and Accountability** — providing a data, algorithmic, and computing foundation that is transparent, accessible, and reinforces accountability across the entire AI value chain (i.e., software engineers, data analysts, users, etc.).
- **Creativity and Ownership** — being able to clearly differentiate data and information owners from users, including the provision of rights to creators of any kind.

- **Social Manipulation and Misinformation** — working to minimize the deliberate misuse of AI for misguided, nefarious, or criminal activities through social engineering, manipulation, or the deliberate spread of misinformation.
- **Privacy, Security, and Surveillance** — providing a foundation that proactively establishes governing rules and procedures that protect privacy and provide the necessary security for AI users; and also protects people against obtrusive surveillance.
- **Job Displacement** — recognition that optimization of a data-driven society will likely result in a shift in workforce needs, rendering some existing jobs obsolete, while also giving rise to new types of career pathways.
- **Autonomous Weaponry** — deterrence of AI-based weapon systems that pose a catastrophic risk and threat to civilians and civil society at large.

It is likely that, as AI advances, it will also give rise to many other moral dilemmas and ethical concerns. Each one of these ethical concerns is important individually, but collectively, they also portray the grand challenge associated with integrating ethical principles into all facets of human-designed technology, infrastructure, products, and services. Each ethical concern feeds into or is connected to another. The interactive effects between these concerns cannot be understated. While our human brains are impressive, we simply cannot predict, plan, or prepare for the infinite outcomes and scenarios that could be derived from the mass deployment of AI into our world. Ironically, AI can be a useful tool and aid in helping us navigate the inclusion of AI into our future. But we must be mindful that AI is a powerful tool and consider the fact that we could even be manipulated in the future by the AI we created. Hence, we must ensure that the "fox is not watching the henhouse," so to speak.

The rapid ascent of AI has grossly underestimated the ethical contract that humans need to embed within the technology and have the means and mechanisms to revisit that contract regularly, to ensure that the evolution of AI is holding up to the values of society in any given instance. We are simply not there yet. AI and the development and assimilation of advanced technology into human and planetary workstreams are deeply in need of pragmatism. AI holds great promise for our future, but its swift advance between 2022 and 2024 was stimulated in part by financial motivation. As we all know, when the allure of quick money is prioritized over longer-term value creation, shortcuts are made, and mistakes can happen. As a case in point, the leadership, manufacturing, aircraft safety issues and failures, and other business challenges faced by Boeing in recent months and years all share money (revenue, earnings, and profitability) as a common denominator. Everyone understands that companies need to be profitable to be sustainable (both financially and in the broader sense of environmental and social sustainability).

Human nature has taught us time and again that, even with the best of intentions, humans built and operated systems can and will fail. With AI, we have an opportunity and a social obligation to identify and prioritize risks and establish an ethical framework and foundation that ensures there is a "human-in-the-loop," figuratively and literally, so that we can enthusiastically yet cautiously work with AI toward the betterment of humanity and the planet.

The New Economy Leaders' Toolkit: An Applied Adaptive Strategy Incorporating the Meta-Dimensions of Sustainability

But Mousie, thou art no thy-lane,
In proving foresight may be vain:
The best laid schemes o' Mice an' Men
Gang aft agley,
An' lea'e us nought but grief an' pain,
For promis'd joy!

From the poem, "To a Mouse," by Robert Burns[65]

In 1785, Scottish poet and lyricist Robert Burns[66] wrote a poem, "To a Mouse," that captures one of the great paradoxes of life, which simply says, "even the best-laid plans sometimes go wrong." Most everyone has experienced the unfortunate and occasionally uncomfortable and even unsettling reality of life, which is, no matter how much we plan and prepare, we cannot control everything to our desire. This is why the insurance industry exists: to manage the risk of the plan (the known) and uncertainty of life (the unknown). It's also why we desperately try to predict and forecast the future — whether it's weather, stock performance, national elections, sporting events, or a friendly wager to determine if our niece will be having a boy or a girl. Humans desperately try to forecast our future so that we can protect and prosper. Knowing this contradiction of human consciousness, we try to hedge our futures by collecting and analyzing data, *a lot of it*. We then try to make sense of the data and build AI models to help us predict the future so that we may be more apt to be proactive and pivot, if necessary. Underlying all our daily attempts to digitize and evaluate life are the *mental models* that support our beliefs, values, and behaviors.

For humans to be adaptive and sustainable, however, we must question the very construct of the mental models, our thinking, and behaviors that uphold our beliefs and which got humanity to its current state of being. Essentially, we must ask and reconcile, did (and do) we have the right plan to begin with? Is the plan incomplete, flawed, or outdated? Might our mental models be limited, biased, incomplete, or flawed? Although we may have tried to have an inclusive, logical, and realistic plan, perhaps we've inadvertently left something critical out? Have we evolved, or remained entrenched and stagnant in our posture toward living peacefully with each other and the planet? For example, no matter how much we integrate active sensors into our cyber-physical world to further digitize and interconnect nature and humanity, or how much data we collect and analyze through sophisticated algorithms, the data and its output will only help us make better decisions insofar as we have principled leadership that establishes the right foundation for strategic planning and pragmatic decision-making in place.

Now, let's consider the foundation for sustainability, which includes long-term and holistic thinking, multidimensional and interdisciplinary collaboration. At their core, sustainability practitioners love to plan! Short-term, mid-term, long-term — strategic and action-oriented planning gets to the heart of how sustainability professionals think, collaborate, learn, grow, and lead. In doing so, they love to put into place the continual improvement plan-do-check-act (PDCA)[67] process as well. Much of a sustainability practitioner's focus and responsibility is placed on establishing a baseline, performance targets, and collecting data to measure and report-out against the baseline and enterprise goals. For example, a company might put the following over-simplified stepped plan into place as one of their sustainability performance pathways:

- *First, Plan* — Establish a greenhouse gas emission inventory based upon a specific baseline year. Then define a performance target such as a 25% reduction by a specific future date.
- *Second, Do* — Put into place an action plan including the resources and strategies necessary to reduce emissions.
- *Third, Check* — Conduct an annual accounting of the greenhouse gas emissions, measuring them against the baseline year, the prior year of measurement and performance, and the future target to assess the enterprise's progress toward its action plan.
- *Fourth, Act* — Evaluate whether the enterprise's capital resource allocation and action plan has achieved the desired result, and then adjust the plan and resource requirements accordingly.

While simplistic, PDCA can effectively aid individuals, teams, and organizations to systematically think through all the necessary requirements for any given problem. Like any framework, there are limitations to PDCA, particularly in measuring the outputs, outcomes, and impact of the acted upon decisions.

Throughout my career, I've always been keen on Impact Models[68] and the Theory of Change[69] model. Simply defined, the Theory of Change framework looks at *how we can affect change* through the overarching formal relationships that are presumed to exist, which impact a defined population. A logic model, on the other hand, is a tool that is used to explore *what we do and how we do it.*

According to the United Nations Development Assistance Framework (UNDAF), Theory of Change refers to "a method that explains how a given intervention, or set of interventions, is/are expected to lead to a specific development change, drawing on a causal analysis based on available evidence." Further, the Center for Theory of Change states that Theory of Change[70] "is essentially a comprehensive description and illustration of how and why a desired change is expected to happen in a particular context. It is focused on mapping out or "filling in" what has been described as the "missing middle" between what a program or change initiative does (its activities or interventions) and how these lead to desired goals being achieved. It does this by first identifying the desired long-term goals and then working back from these to identify all the conditions (outcomes) that must be in place (and how these relate to one another causally) for the goals to occur. These are all mapped out in an Outcomes Framework."

Theory of Change and Logic Models are great methods to deploy, especially when diverse groups and project stakeholders seek a logical process to define the causal ties and relationships between the inputs and resources needed to plan, execute, and derive specific impacts across a defined time horizon, usually a longer-term one. Essentially, these methods help stakeholders engage on the "what and how" of putting resources into action to achieve a shared result. I was exposed to these methods very early in my career when I worked as an energy analyst on the program evaluation team for the New York State Energy Research and Development Authority (NYSERDA).

Residing within NYSERDA's Energy Analysis group at the time, the program evaluation team was charged with the independent evaluation of NYSERDA's public benefit program that included energy efficiency, renewable energy, energy research and development, and other programs and initiatives that leveraged New York State electric-and-gas utility ratepayer funding toward the design and implementation of energy-related public benefit programs. Our program evaluation team used Theory of Change and Logic Model methods to support strategic program planning and evaluation, helping to ensure that the ratepayer funding was deployed responsibly and equitably, and with its intended efficacy. Working with external experts and an amazing team of professionals, I quickly learned the value of these methods for stakeholder engagement, program design and management, strategic planning, and organizational change management. This early career exposure to program evaluation sparked a skillset and a desire for "applied strategy," utilizing the Theory of Change and Logic Model methods to define and measure organizational performance metrics not only as a scorecard, but also as a mechanism to ask deeper and more informed questions and a means to engage stakeholders thoughtfully and intentionally so that all parties could derive better programs and results.

My NYSERDA experience was the beginning of a career spent going deeper into the underlying values, purpose, and "who and why" of organizations and people. It's led me to understand that humans absolutely love collecting data, formulating, and operationalizing processes, and measuring the performance of people, business systems, and operations. These are all critical components of operating any enterprise effectively. Knowing what data to collect and how to analyze and synthesize it into meaningful indicators of performance is essential to enterprise success. However, throughout my career, I have witnessed time and again that humans love data, processes, and performance measurement so much that for some, these become the walls and canopies that give them cover.

I've also learned that any individual and organization is prone to having bias and blind spots when evaluating data; even when best-in-class evaluation frameworks, tools, software, and team structures are deployed. Whether intentional or not, people (and organizations) tend to hide behind data, processes, and the mechanism of reporting out on their activities. Let's face it, in business, this gives people something to measure, say, talk about, and improve upon. It can be easy and convenient to "own a process," "manage the data," and "hold others accountable to their performance."

Interestingly, the Logic Model and the Theory of Change methods get at two of the three elements of what author and speaker Simon Sinek calls the "Golden Circle," as referenced in his 2009 book, *Start with Why: How Great Leaders Inspire Everyone to Take Action.* The Logic Model and the Theory of Change methods get to the *how* and the *what,* but they don't sufficiently get to the underlying why, that is, the true purpose behind the enterprise, culture and motivation for being. If the underlying foundation and basic principles for *why* the enterprise does these things is not tethered to strong ethics, values, morals and conviction, well then that is the recipe for the status-quo kind of culture and environment that doesn't see the iceberg ahead in the fog lights, let alone have the level of leadership (the who) that has the gumption and accountability to let the crew know to steer clear.

A plan alone is not enough, and too often, even our best-laid plans go awry. As such, we must remain open to anything that may positively improve upon the design and implementation of the strategic plan. For example, being able to scan the horizon and monitor trends or drivers that influence a business enterprise (or one's own life) is a great skill to develop. Forecasting the future with precision that brings a great deal of certainty is challenging. The best financial market, weather, sports, political, and consumer trends forecasting modeling software and analysts have a margin of error.

Forecasting for a general sense of the "market condition" or to assess "which way the wind is blowing" can prove to be a useful tool to an organization seeking to define its roadmap for X (with X being innovation, growth, sustainability, profitability, and so on).

Forecasting very granular impacts across financial, people, operational, or innovation performance is never precise. Therefore, the best organizations evaluate a range of scenarios, margins of error, and assumptions so that a range of possible outcomes and impacts are evaluated. This approach prepares and conditions the organization and its leadership to be open to change, remain adaptive, and to preposition its mindset and resources for any scenario. In this way, "applied adaptive strategy" is an approach that provides clarity on direction and resources and flexibility on allocation and focus. Although this may seem to be a contradiction, it provides the underlying requirements for how organizations compete in a dynamic, uncertain, and risk-based economy. There are several ways to adopt an applied adaptive strategy to an enterprise. Whether your objective is enterprise growth and innovation, strategic planning, risk management, sustainable impact, or other priorities, the tools and frameworks already mentioned offer a great starting point for planning, implementation, and performance measurement. There are also many other strategic management framework tools including the well-known SWOT and sSWOT (Strengths, Weaknesses, Opportunities and Threats; and the "s" for, Sustainability SWOT), the Business Model Canvas (BMC), the World Resources Institute's (WRI) Connection Matrix, PESTEL Analysis (Political, Economic, Social, Technological, Legal, and Environmental), and the Value Chain Analysis conceived by Harvard Business School Professor, Michael Porter.

Each framework tool has a specific purpose. Organizations that proactively manage risk and change integrate multiple framework tools together to provide a more comprehensive assessment of the known and unknown factors that are or that may influence the business direction, resources, and performance. I'm a huge fan and believer in applied adaptive strategy that integrates multiple strategic management framework tools into a pragmatic roadmap for the enterprise. Simply defined, a roadmap provides the necessity for having a plan of action that is disciplined, yet adaptive and incorporates the dimensions of change, time, resilience, and human ingenuity. In doing so, the roadmap inherently builds in internal and external inputs, capturing insights from the business and its diverse mosaic of stakeholders.

Incorporate the Meta-Dimensions of Sustainability into Your Adaptive Strategy Process

A couple of years ago, I gave a talk at Wells College in Upstate New York titled, "Discovering Your Future in the Meta Dimensions of Sustainability[71]." My purpose for this talk was to provide dynamic discussion on *"current affairs, the state of sustainability in society and business, the role of individuals-citizens-consumers, and the nexus between corporations, government, and society on truly manifesting a sustainable future. Many have a contextual sense of sustainability, but sometimes they don't fully see or understand the complexity of systems or the nuance of different points of view and sectors. The sense of urgency on climate action is rapidly intensifying. This can leave many feeling either invigorated or helpless, frustrated, and even apathetic to the state of affairs and fate of their future in the face of climate and sustainability risks. Having perspective on the role one can serve and having a vision for one's future, given the need to change and adapt, can be empowering and uplifting and enable the individual to have resolve in how they engage within the multi-dimensions of sustainability."*

Wow, that is a mouthful! Looking back, I had a great time preparing for this talk. A colleague, Marian Brown, who was the Director of the Center for Sustainability and the Environment at Wells College, had invited me to their Sustainable Business Series as a speaker. I was delighted and had a great time preparing for the Wells College presentation.

As I prepared, as I had done hundreds of times before for academic classes, public meetings, or business events, I wanted to elevate and illuminate a sample of the most pressing converging drivers impacting business and society for discussion. For fun, and using alliteration of the letter "D," I devised "seven meta dimensions" of sustainability as the central theme for the talk and discussion.

A Fifth Wave of Social Transformation?

How Meta-Dimensions of Sustainability are Shaping a New Generation of Leaders

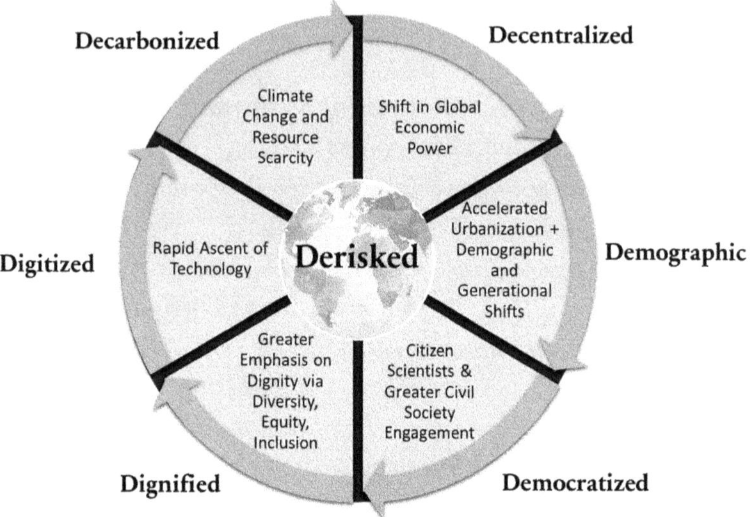

The economy (and greater society) are undergoing a continuous, albeit an increasingly more rapid state, of change. The convergence of sustainability megatrends adds complexity – both challenge and new opportunity – to the current and future state of how business continues to create value.From my experience over the past twenty years, and certainly in the past five years, the *following seven drivers* have added more depth and meaning to the evolution of sustainability in popular culture, government, business, and civil society. Each dimension stands on its own as a mega force for sustainability, but each of these is also interlocked, contributing to network effects. The more people interact with these drivers, individually and collectively, the greater the awareness and likelihood of sustainability's advance in business and society.

Seven Meta Dimensions of Sustainability

Dimension	Description
Demographic	The world's population is shifting before our eyes. The United Nations predicts that by 2050, greater than 50% of the world's population will live in urban areas[72]. The population in developing countries skews younger than developed countries. Populations in less developed countries are expected to double between 2022 and 2050, whereas the population in many developed countries is projected to remain flat or decline. Demographic shifts toward greater urban density and higher population growth rates in less developed countries will place new demands on resource consumption for basic needs and economic growth. Demand for energy, clean water and sanitation, food commodities, healthcare, safety, and security is projected to rise for urban areas and within developing countries.
Democratized	Democratization has been a force that has been evolving steadily in recent years. Even amid the ugliness and backdrop of demagogue politics, there has been a persistent counter movement focused on diversity, equity, and inclusion (DEI), environmental and social justice, and the opportunity for citizens and consumers to have greater choice and say in what shapes their quality of life and futures. Inherent within democratization are diplomatic strategies and solutions. The art of stakeholder engagement requires the ability to convene and foster dialogue across a diversity of people and their viewpoints. Facilitation requires active listening skills and an ability to remain independent, objective, and diplomatic. In recent years, as demagogue politicians have fueled the fire of polarized viewpoints, diplomacy seems to have waned. A push for pragmatic, commonsense, diplomatic discourse is underway countering the less civil tone that has provided entertainment value, but not much more.

Dimension	Description
Decentralized	Continued advances in technology are partially stimulated by the decentralization of formerly centralized systems. Examples include finance and investing (i.e., decentralized finance, or DeFi, including cryptocurrency and blockchain), energy and infrastructure (i.e., renewables, microgrids), transportation (i.e., autonomy, EVs), education, healthcare, government services, and cybersecurity. Our data-driven world, reinforced by new technology that promotes transparency, accessibility, and inclusivity, is placing more power and control in the hands of citizens and users and is driving a fundamental paradigm shift toward decentralized business models.
Decarbonized	The accounting, management, reduction, and elimination of excess carbon emissions is arguably the most popular and pressing concern for global politicians, corporations, and governments. Decarbonization is like the Taylor Swift of sustainability objectives. Decarbonization is iconic and classic. It's edgy but not so much that the "old dogs" won't join in for some fun. Carbon (accounting and management) has been on tour for a long, long time, like the Rolling Stones. It's still incredibly relevant and gets the old-time companies "off their feet," while still attracting some new fans. Carbon (old) and decarbonization (new) take center stage at most global sustainability conferences. It's like Mick Jagger strutting, and Taylor Swift gracefully gliding onto stage to wow fans — that's Carbon and Decarbonization in action. It's an awesome sight, and you know it's going to rock. If we tallied the value of investment that has already and will continue to flow into decarbonization technologies and pathways between now and 2030, it certainly would take top billing compared to other sustainability goals. Like The Stones' and Swifties' fanbases, pathways for decarbonization have proliferated, offering yet another paradigm shift in how humanity thinks about its relationship with fossil-based fuels and their emissions, namely carbon dioxide.

Dimension	Description
Digitized	Mathematician Clive Humby[73] has been credited with the phrase, "data is the new oil." Although his connotation was directed toward creating value from data, just as oil is only valuable insofar as the molecules are refined into useful fuel, chemical, lubricant, or plastic, the quote speaks volumes on our current reliance on digitizing everything into data. As a societal driver, however, digitization is having an enormous impact. On one hand, our data-driven digital society has given exponential rise to power-hungry datacenters, the networked servers and supercomputers that transact all the data that is collected, analyzed, and used to manage our digital society.
Dignified	Take a review of the United Nations' 17 Sustainable Development Goals[74] (SDGs) and you'll see that more than half of the SDGs are squarely aligned to social impact and the people side of sustainability. For example, SDG1: No Poverty, SDG2: Zero Hunger, SDG3: Good Health and Well-Being, SDG4: Quality Education, SDG5: Gender Equality, SDG8: Decent Work and Economic Growth, SDG10: Reduced Inequalities, and SDG16: Peace, Justice and Strong Institutions. The UN's SDGs are but one example of a global sustainability framework that is working toward aligning the dignity of all living things. Ensuring the dignity of all peoples and cultures has become a common refrain, taking on a more central priority for government, industry, and civil society in recent years.

Dimension	Description
De-Risked	Although de-risking anything may be construed as a logical objective (i.e., derisking business, the economy, your healthcare options, the purchase of your next home, etc.), the sustainability space is exploring derisking from multiple pathways: (1) derisking the investment into sustainable technologies including clean energy, (2) derisking the implementation of the "sustainable solution," (3) derisking business and societal impact on the environment, (4) derisking future climate risks through climate adaptation and mitigation strategies, and more. Risk management has always been a key component of environmental management; however, it is now also fully intertwined within the broader sustainability agenda. The inclusion of risk management into social and governance aspects of ESG, for example, supports the intention for more holistic systems-level solutions for business, government, and society. Derisking sustainability is a necessity. There is a need for humanity to be preventive, predictive, and proactive in our posture toward our relationship with each other and the planet so that we can alleviate the prolongation of existing or proliferation of new, unintended consequences of our actions. We now have the knowledge, technology, and capacity to live in greater harmony with nature and each other. With a principled foundation in place and the will to serve and act, humanity can derisk our future and that of the next generations of people from undue harm.

Wait, the "Seven-Ds" are Incomplete

In November 2022, I had the privilege to attend the Railroad Environmental Conference (RREC) at the University of Illinois at Urbana-Champaign and facilitate a panel of Class 1 railroad sustainability leaders. The panel was made up of representatives from the Association of American Railroads, BNSF Railway, Canadian

National Railway, CSX, Canadian Pacific, Kansas City Southern Railway, Norfolk Southern, and Union Pacific Railroad. Our discussion centered on "Sustainability and Resilience: Driving ESG Impact through Supplier Engagement." In support of how the panel members would segue into their remarks and the discussion, I referenced the "seven meta dimensions" of sustainability in my opening remarks to establish a foundation. The strategy worked, and a great conversation ensued.

At the end of our session, we fielded some questions from the audience. One of the participants approached the microphone and praised the panel. He also provided feedback on the seven meta dimensions, suggesting that there was one missing, an eighth dimension he called "decisiveness and decision-ready doers." The participant noted that another significant driver in the sustainability domain is in making the informed DECISION to act. He pointed out that sustainability practitioners ultimately have a responsibility to take all the drivers, assess their opportunities and threats, and eventually prioritize their goals, strategies, and resources into decisive action. I welcomed and loved the point this gentleman made and decided to add an eighth "D" to the meta dimensions to account for the uptick in more sustainability-focused decisions, leading toward impact.

As a ninth "D" option, I'm also adding "diversified" or "diversification" to the meta dimensions list. You've likely heard the phrase, "there is no silver bullet" technology, product, or service that can singularly "solve for X." In fact, many of the environmental, social, and economic challenges that plague us today are the result of us "going heavy" on one type of energy source (i.e., oil and energy price volatility and climate risk), or food source (i.e., beef and loss of old growth forests and carbon dioxide sinks, and elevated occurrences of heart disease), or transportation modality (i.e., the car culture in the U.S. vs. the more efficient utilization of high-speed rail and mass transit in other nations). In these examples, and many others, we have discovered that greater efficiency, resilience, and quality of life can be attained when a diversity of options for "solving for X" exist. This is why the energy, infrastructure, transportation, and agriculture sectors are exploring diversification to secure their future. Diversification of natural resources, people and talent, data and information, investments and ideas all support planet prosperity and our future sustainability as a hedge to risk, a reinforcement for resilience, and enabling innovation to "solve for X" with greater efficacy and potential for success.

Although the alliteration of the letter "D" is fun and memorable, there likely are many other "Ds" that are in development or currently left out of the equation. This is the larger point in my view, however. Just as the audience member suggested something new to advance a loosely held together framework of industry drivers, the sustainability movement, skillset, and opportunity for creating an impact are *ever-changing and evolving.* The point of the seven, now "nine-Ds," is to continually be present and aware of the world around us. Take in the signals that point toward change and envelop the right data and principles to guide smart and practical decisions.

Foretelling the Future Takes Discipline

The building blocks for strategic planning and adaptive roadmaps are drawn from many internal and external resources. Linking existing trends (i.e., exploring those elements currently shaping the economy and enterprise landscape) with future casting (i.e., leveraging market, business, technology, and customer intelligence to foretell a range of future-based scenarios across specific criteria for estimating probability and likelihood of change) provides a reasonable sense of future direction.

*Special note: In my experience, the use of strategic management framework tools, applied adaptive strategy, and roadmaps should be valued as an evergreen process. This can drive some organizations mad, however. Business leaders like certainty and prefer to have clear targets and well-defined intervals for business planning and resource allocation. Business leaders want a level of predictability for performance measures and certainty of outcomes. The reality, however, is that business operates in a dynamic and ever-changing global environment.

Strategic plans should not be so rigid that they are inflexible and become a liability to the enterprise. Change is inevitable; it happens. Business cycles tend to operate in quarters, and too often, enterprise processes mirror this cadence. Milestone reporting periods are just that, slices of time where aggregated performance is measured against a benchmark. Those timestamp reporting periods provide useful measures of business performance against the strategy; but it is the adjustment of strategy, thoughtfully and continuously throughout the cycle, that ultimately determines how calibrated the business will be. From another perspective, business strategy is never a "one and done" effort; it is an everyday immersion into the assessment and evaluation of the forces that shape the business's decisions.

An applied adaptive strategy can produce useful products, like a roadmap or report that lays out the direction for the business. The real value, however, is in its ability to provide the business culture, leadership, and teams with the necessary insight to be resolute in their adaptation of the plan at any given moment.

Managing change requires the following[75]:

- A Proactive Applied Adaptive Enterprise Strategy
- A Sense of Urgency: Agility + Discipline
- A Multi-Stakeholder Leadership Coalition with Firm Commitments
- Pragmatic Vision Framing, Roadmapping, Measurement, and Refinement
- Organizational Readiness & Redefinition

Getting the most out of an applied adaptive strategy requires ***ongoing, clear, and inclusive communication across the enterprise.*** Applied adaptive strategy is not (exclusively) about the roadmap, but ultimately about who is taking responsibility to ensure the organization is always alert and ready to adapt, learn, relate, and respond to change. Deployment of an applied adaptive strategy that includes the meta-dimensions of sustainability as a means for continuous monitoring and feedback to leadership teams throughout the business cycle can lead to greater stakeholder awareness (and engagement) and a more robust analysis of scenarios for growth.

Your Feedback and Participation in Leadership are Essential for Enterprise Success

Too often, business strategy is thought of as a "black box," where few people know what's inside or how the inner workings produce useful outputs and insights. When executed well, an applied adaptive strategy provides a means to engage stakeholders within and across the enterprise, with use of structured frameworks and open-ended inquiry, that supports a more diverse and inclusive participation. The net result is a broader, deeper, and more holistic view of business drivers, risks, and opportunities, enabling the enterprise to establish adaptive scenarios that factor in the meta-dimensions of sustainability and change.

What additional "meta-dimensions" (i.e., macro-level drivers of change and influence upon the enterprise, markets, and the economy) do you see actively shaping the current and future state of sustainable enterprise and sustainability within society? How is your organization taking stock of these meta-dimensions and incorporating them into your short-to-long-term strategic planning? Who within your organization is charged with assessing your organizational strengths and weaknesses against these drivers and influences that pose both opportunity and threat to the enterprise? How are stakeholders, including employees, vendors and suppliers, customers, government regulators, market allies, investors and shareholders, and members of the board, interfacing with these meta-dimensions of change? How is the organization remaining agile, adaptive, and inventive in its posture to interpret and act on these and other market-shaping meta-dimensions?

Principles for a Pragmatic Planet

A strong dose of *realism*. An infusion of *common sense*. A swift and piercing shot in the arm of *pragmatism*. However you want to take it, that's what is needed right now, in this moment of global generational and demographic change. If you have not yet noticed, a lot has changed in the past few years. Some might say, everything is different. Some might say that the speed of change is welcomed. Others might say, let's pump the brakes, hard, come to a complete stop, and turn this car around! And yet others are burnt-out on all the supercharged rhetoric and have grown, much like a stereotypical Gen-Xer, apathetic to the state of change and affairs that are actively shaping our world. I sense and believe wholeheartedly that people are sick and tired of extreme personalities, points of view, and politics.

Think of the Earth and humanity swinging like a pendulum between the polar extremism of the far left and right, as shown below. The pendulum of extremism has been steadily swinging between the far right and far left of logical thinking and pragmatic behavior for the past few years. There are clear and present risks on either side of the far left and right. There is also a zone of pragmatism, closer to the center, "where things can get done." The zone of pragmatism is not devoid of risk. It takes principled leadership to effectively navigate change while making sound decisions. The zone of pragmatism bridges the deep chasm that has created a barrier between citizens, consumers, and all stakeholders.

Acting from a zone of pragmatism enables us to reintroduce common sense for the common good into the fabric of our relations. In doing so, we can have more constructive dialogue that yields decisions that are more informed, principled, and pragmatic for people and the planet. For the past few years, the pendulum of extremism has swung rapidly and even violently between the polar right and left. This has created and fueled tension, uncertainty, fear, and cynicism among citizens and consumers. Overlay this period of divergent and divisive discourse with economic insecurity and social unrest, and people begin to feel as if their American promise of prosperity has been lost.

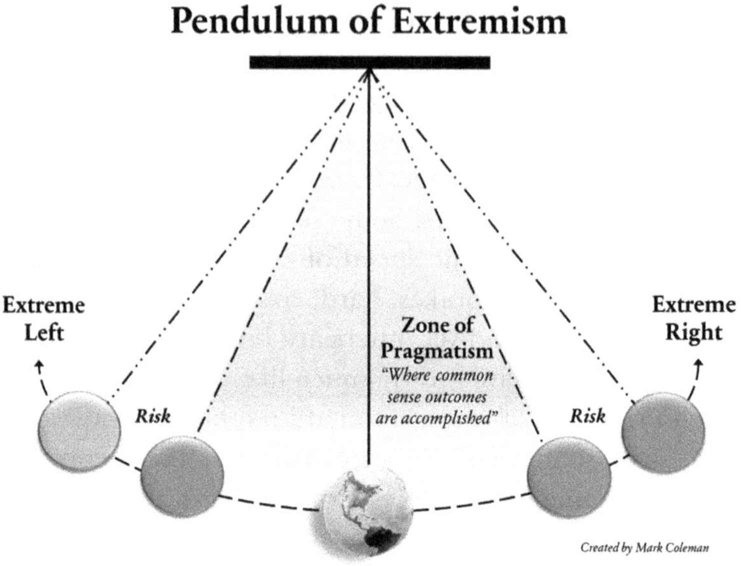

Pendulum of Extremism

Extreme Left

Risk

Zone of Pragmatism
"Where common sense outcomes are accomplished"

Risk

Extreme Right

Created by Mark Coleman

Many of us have been passively watching and listening from the sidelines, as the entertainment of extremism has intercepted our lives with new names, faces, and perspectives on all that is going awry in the world. It has been frustrating, disheartening, and annoying to witness the proliferation of people with polarizing principles ratcheting up their entrenched positions, often with extreme behavior. This moment of derelict extremists cooking up distrust and divisiveness in American politics and culture is a broader mirror to our global society. The past few years of extremism may have been a needed release for a wider population to see, hear, and try to understand the realities of the world around them.

I'm going to work hard and try not to walk you, the reader, through all the historic events and societal pain points, or the twisted labyrinth of nonsensical lies we've been telling ourselves, or which have been told to us as associated with demagogue politicians, unsocial media, entertainment news pundits and sources, or "newly informed" neighbors, friends, or family members that led to our wildly swinging pendulum. Doing so would be a lose-lose endeavor. If you're reading this, then you've been alive and have formulated your perspective on how the world around you is unfolding and what it means for your life. My objective is not to sway your pendulum of perspective in one direction or another. I do hope, however, that we can agree that extremism is not leading any of us to a better quality of life or to greater prosperity. In fact, it's wasting our time, energy, and patience.

Extremism has led us down a demonic path of unproductive self-righteousness. In this talk louder, act bigger and bolder, bullyish climate of nonsensical debate, everyone is losing. We all know that "the squeaky wheel gets the grease," and certain demagogues in training have made it their selfish mission to say and be as outrageous as they possibly can. They know that the media loves soundbites, and that most Americans have short attention spans and are easily entertained. Thus, these dissidents deliberately seek to obfuscate reality, distort the truth, and try to be as squeaky as they can be. Looking back at the disdain and distrust exhibited by so many people over the past five to eight years, it's no wonder we continue to see an escalation in mental health, civil unrest, geopolitical dysfunction, and social upheaval. People have been generally unloved, fearful, angry, alone, hungry, and in search of truth and justice.

Peace, prosperity, and the planet lie in the balance. In recent years, we've witnessed a deterioration of social institutions that have traditionally provided the glue that binds people, ideas, and political will together. As social institutions and constructs have been turned upside down, it has created a vacuum that has been filled with an endless barrage of sensationalized soundbites that have fueled anger, apathy, and distrust. People are generally less tolerant and trusting of each other. When social glue no longer bonds us together, we falter. As a result, democracy is weakened.

Unfortunately, "we the people" have turned on each other, more than any time in my living memory. We currently lack the necessary leadership, common purpose, and common sense in how we are pursuing what I call planet pragmatism — which is a means to say — prosperity for all living things, people and planet. Plain and simple, people are in search of principled leadership and a new pathway to prosperity.

This book was conceived with the original thought of providing us with a playbook for getting back to the center, back to basics. I believe people are in pursuit of a more principled, pragmatic, and purposeful future. Accordingly, I believe people are yearning for the same traits in their leaders across the spectrum of stakeholders that occupy their lives, from politicians and elected officials, corporations and governments, to community-based organizations and our public safety and service professionals.

As we get started, let's explore a couple of foundational definitions for wisdom and pragmatism:

- ***Wisdom*** is defined as the "ability to think and act using knowledge, experience, understanding, common sense, and insight."
- ***Pragmatism*** refers to "an approach that evaluates theories or beliefs in terms of the success of their practical application."
- Principles of pragmatism[76] are (1) **an emphasis on actionable knowledge, (2) recognition of the interconnectedness between experience, knowing, and acting, and (3) a view of inquiry as an experiential process.**

We live in a world of irony and paradox marked by contradiction, inconsistency, and ambiguity. Consider phrases we use such as "sustainable growth," "radical innovation," "artificial intelligence," and "clean energy." Irony is a part of our human existence. I'm reminded of the popular singer and songwriter Dave Matthews and his band's song, "Funny the Way It Is" from their 2009 Big Whiskey and the GrooGrux King album, which points out some of life's ironic contradictions.

As depicted in Dave Matthews's song, we live within and navigate a daily dose of contrasts. We live in a society where extreme wealth and poverty can co-exist on the same street in Manhattan, Silicon Valley, or Washington D.C. (i.e., the financial, technology, and public policy capitals of the U.S., arguably the world).

It can be jarring, for example, to witness the extreme and stark contrast of clearcutting of tropical rainforests to harvest ancient wood and make room for agriculture production. In April 2024, New York State Attorney General Letitia James's office filed a lawsuit against JBS, a Brazilian multinational corporation that is the world's biggest meat company. In 2021, JBS began to proclaim that it would achieve net-zero emissions by 2040. JBS launched their climate campaign in 2021, where in big, bold font on a full-page New York Times[77] ad that read, "Agriculture Can Be Part of the Climate Solutions." Following this attention grabber, the advertisement went on to make the statement, "Bacon, Chicken Wings and Steak With Net-Zero Emissions. It's Possible." Also in 2021, JBS reported that it had greater than 71 million tons of carbon dioxide equivalent (CO2e). For context, according to the New York State Attorney General's office, JBS's global carbon emissions are greater than the entire country of Ireland[78].

As global demand for beef and meat products has increased, companies in the agriculture and meat industry have expanded their operations into ecologically sensitive regions of the world, including the Brazilian rainforest. When precious old-growth forests are clear-cut to optimize for agriculture production, the immediate and long-term impacts on key ecosystem services, including the carbon and water cycles, are exacerbated. The United States Department of Agriculture (USDA) website[79] defines ecosystem services as "the direct and indirect benefits that ecosystems provide humans. Agroecosystems, rangelands, and forests provide suites of ecosystem services that support and sustain human livelihoods." The USDA goes on to add that there are four areas that ecosystem services are typically broken down into, as noted below.

Four Categories of Ecosystem Services
Source: US Department of Agriculture, "Ecosystem Services"

- **Provisioning services:** the material or energy outputs from an ecosystem, including food, forage, fiber, fresh water, and other resources
- **Regulating services**: benefits obtained through moderation or control of ecosystem processes, including regulation of local climate, air, or soil quality; carbon sequestration; flood, erosion, or disease control; and pollination
- **Supporting services**: services that maintain fundamental ecosystem processes, such as habitat for plants and wildlife, or the maintenance of genetic and biological diversity
- **Cultural services**: the non-material benefits that ecosystems provide to human societies and culture, including opportunities for recreation, tourism, aesthetic or artistic appreciation, and spirituality

Rainforests provide several critical ecosystem services, including soil formation and nutrient cycling, climate and flood regulation, water purification, and the provisioning of life-essential resources to local peoples, including food, wood and fiber, fuel, and fresh water. Further, ecosystems are also culturally significant in their role serving local people and communities with spiritual, educational, recreational, tourism, and aesthetic benefits.

At this time, it is uncertain whether JBS will face any financial penalty associated with the New York Attorney General's lawsuit. The lawsuit raises awareness of JBS's environmental claims and the overarching impact they and the global agriculture and meat industry have on global carbon emissions and the deforestation practices associated with expanding grazing lands to support cattle ranching.

The challenge humans have had, and will continue to have in the foreseeable future, is that sustainability is a pursuit, not a destination. It is one thing for humans to define and place sustainability metrics on a singular industrial process, consumer product, or global corporation. It's an entirely different task to measure, quantify, and control planetary processes. Our existence with the planet is an input-output model, thus conceivably, we can control our environmental footprint and impact by simply limiting any negative intake (i.e., resource extraction) and negative output (i.e., waste, emission, inefficiency, etc.). Our existence is much more nuanced than this, as is our scale. With eight billion people living in 195 countries and growing, humans have created and used a lot of input-output models to assess, measure, refine and update the systems that serve our life (i.e., financial, energy, agriculture/food, transportation, education, healthcare, government and defense, space exploration, research, and innovation). We know that resource efficiency and optimization do not necessarily yield sustainability either. We have other influences that shape our need to survive and desire to thrive.

The planets' ecosystems are alive and dynamic. The planet is ever-changing, modulating on short- and long-term horizons the capacity to continuously deliver the four categories of ecosystem services to us. Some of our input-output activities have been created to support nature and ecosystems, such as afforestation, wetland mitigation, and other nature-based solutions. As a result, our input-output models have become more complex, and some might even argue that there is evidence to support the idea that just as we need nature, it also needs us in certain circumstances. Humans are now working to research and understand how we can mitigate the loss of coral reefs, including replanting corals and restimulating these life-critical ecosystems that have been on the decline. In this way, humans and nature have never been separated from each other. As humans come to appreciate and understand that the planet and our existence are intrinsically linked, the context, purpose, and potential for sustainability will become clearer to the masses.

We are not there yet. Humans are mighty in number, illuminated by our mass consumption, waste, and relative destructiveness to each other and the planet. We remain, however, immature in harnessing our collective wisdom, much of which has been born from nature, to live in greater respect and harmony with each other and the planet. As such, the pursuit of sustainability is an enigma and can certainly be elusive. At the heart of all sustainability questions is a recurrent theme and reality, which is the fact that humans need the environment to survive. We need the planet for survival. Although we work hard to design and engineer a human-built world that provides for our own comfort and well-being, we are always having to contend with the realities of a dynamic and continuously changing planet.

The planet is alive, and even aside from our influence, it constantly changes. The air, water, land, fisheries, animals, and plants are all in flux at any given moment. There never has been, nor will there ever be, a true steady state for the planet. To minimize our risks and optimize our quality of life, we attempt to control the environment. But we know that these attempts can be short-lived and even futile in some environments, or without the continual reliance on a steady supply of resources. If you think about it, a significant portion of our carbon footprint and sustainability challenges stems from our "industrialization of survival," that is, our innate desire and need to provide the necessary provisions for survival: food, shelter, clothing. Ironically, through unsustainable industrial production and consumption, we have and continue to diminish the clean air, water, and soil quality in the process, a self-defeating proposition toward long-term sustainability and survivability. Delivering a risk-free, safe environment for all humans does not equate to sustainability. Sustainability, then, is about doing the best we can, while we can, at any given moment of existence.

College-level courses like Environmental Studies 101 often teach that human population growth is one of the main culprits of uncivil, unjust, and unsustainable outcomes throughout the world. I've raised that point with my undergraduate and graduate sustainability students. The more people the world has, the more mouths there are to feed, the more resources are needed to be consumed, and the more waste is produced. That perspective feels very dated, incomplete, and outright wrong in today's world which is thinking through more circular and sustainable models of production and consumption. I personally believe humans have an incredible capacity to learn and evolve, and in so doing, bring about a higher consciousness and deeper intellect for survival, which we now know requires us to do so within the planetary operating characteristics of today, tomorrow, and for any future generation. The planet is dynamic, and so are we.

Restoring the planet requires us to first restore the principles encoded in the deep knowledge and wisdom that we already have. Humans and nature are not separate; we have always been interwoven as one sanctity of life. Sustainability has no beginning or end. It is an "evergreen" pursuit and will have relevance in every moment and for every generation. That does not skirt our responsibility for having a long-term, holistic viewpoint regarding how our decisions impact future generations. Rather, it requires us to be and live in the moment, but also sense the future and elevate our wisdom and intellect to ensure survival. This will challenge us and future generations to let go of many human-based convictions rooted in egoism, selfishness, and a desire to control everything around us.

The human-conceived ideas of ownership and permanence are enormous biases and blind spots that limit our ability to embrace sustainability and pragmatic principles for the planet more intimately. Sustainability has always been and shall remain a dance between all living things (and perhaps even the unexplainable phenomena that exist within our world and Universe). The dance can be beautiful, exhilarating, emotional, and exhausting. Remaining interlocked in the dance requires us to pay attention to the rhythm, tempo, and movements of our dance partners. To continue to dance, we must see each other. We must respect and trust one another. We must love one another unconditionally. Human occupation of this planet can only happen within this delicate embrace.

Points on Pragmatism

- *Multiple "meta-dimensions" of sustainability are proactively reshaping the world around us. More than data points, indicators, or trends — the meta-dimensions of sustainability represent an underlying and profound shift in how we intentionally pursue a new prosperity, one that innately connects people and planet.*
- *Many of us have been passively watching and listening from the sidelines, as the entertainment of extremism has intercepted our lives with new names, faces, and perspectives on all that is going awry in the world. The past few years of extremism may have been a needed release for a wider population to see, hear, and try to understand the realities of the world around them.*
- *The pendulum of extremism has been steadily swinging between the far right and far left of logical thinking and pragmatic behavior for the past few years. There are clear and present risks on either side of the far left and right. There is also a zone of pragmatism, closer to the center, "where things can get done."*
- *Restoring the planet requires us to first restore the principles encoded in the deep knowledge and wisdom that we already have. Humans and nature are not separate; we have always been interwoven as one sanctity of life. Sustainability has no beginning or end. It is an "evergreen" pursuit and will have relevance in every moment and for every generation.*
- *An applied adaptive strategy can produce useful products, like a roadmap, playbook, or report that lays out the direction for the business. The real value, however, is in its ability to provide the business culture, leadership, and teams with the necessary insight to be resolute in their adaptation of the plan at any given moment.*

10

EXAMINING PLANET PRAGMATISM IN THE DIGITAL AGE

From Ancient Scrolls to Modern Scrolling: Are the Sands of Time Shaping Humanity's Future for the Better?

Did you know that the average human attention span is 8.25 seconds! Humans have shorter attention spans than goldfish (who come in at a whopping 9 seconds!). Wait, what?

According to some sources[80] the average human attention span has decreased nearly 25% between the years 2000 to 2015. By the time you've read these facts, something is likely already competing for your attention. Well, hopefully I didn't lose you to that video of some dude power washing decades of filth off a late 80s Chevy Iroc Z28 found in a rural North Carolina barn. Deceptively satisfying, I know. If you can, stick with me for at least a few more seconds!

Influencers and marketers know that people can only pay attention for short snippets of time. Marketers also know that consumers don't like to retain a lot of information. Accordingly, marketers have long been feeding consumers a steady diet of advertisements that seek to influence behavior, specifically consumption behavior, by communicating in very short, simple, and entertaining ways. The short attention span of consumers has been trained by influencers and advertisers that benefit from the culture of consumption, by which they reinforce. *See, click, buy...repent? Nah,...let's repeat!*

Take a pause and reflect on this for a moment. Our fast-paced, digitally enabled, and enhanced media culture is continuously stimulating our desire for instant gratification. Think of it as a quick hit of caffeine or, better yet, dopamine, every few seconds. Unless we deliberately choose to power down and decouple from the matrix, we're constantly bombarded with images, text, music, aromas, and much more every moment of the day, determined to capture our attention, and intentionally seeking to influence our short- and long-term thoughts and behaviors. If that's not being plugged into the matrix, then I don't know what is.

Listen, I'm not above scrolling. *Believe me, I've been there, and I get it.* I, too, have scrolled and sat captivated by clips, reels, and posts. You know the ones I'm referring to, the clips that lure you in with fashion and fitness (i.e., the clothes that you can't afford, or that elusive body image — "If I just do a few more planks and burpees"), or the reel that cuts to the chase with a six-month home remodeling project condensed to a 30-second reveal (i.e., I need that swanky backyard barbeque and modern kitchen, like now!), or the post appealing to your inner child with a slow-motion camera reel that follows the smooth sexy lines of an exotic car with cool music beats to boot (i.e., hey, a guy can still dream can't he?).

I give a lot of credit to the creators and influencers out there. They are certainly creative and understand how to charm us and capture our attention (and more importantly, our precious time!). I'll admit that scrolling can be entertaining. It can offer a moment of needed, yet temporary reprieve from the responsibilities of modern living. And, on occasion, it can lead to some practical insight or perspective that can be used.

Scrolling can have utility. But if I'm being honest with myself, and we are all being honest with each other, let's face it, scrolling can be an addictive time sink that yields little positive impact on our lives, and comes with an unintentional emotional and psychological price tag. Have you ever walked away from scrolling feeling less than, or worse, anxious, depressed, ashamed, confused, or lonely? The more we bombard our brains with flash in the pan images of perfection, we lose a little bit of our compass bearing on our true north setting, that is, who we are and what we truly desire to be (as opposed to being influenced to be something "better," through the barrage of curated lifestyles created by people who aren't even really living that life).

Unfortunately for the United States, although consumer protections exist, "on the books," they are not as protective or as policed as they are in other countries. The responsibility for providing discerning evaluation and critical judgment on social media influencers and advertising claims, and their messaging, resides largely with us, the consumers.

From another perspective, we've all seen those incessant pharmaceutical commercials that interrupt our favorite programs. I'm willing to wager that you know, by heart, more than one of the pharma drug commercials' catchy tunes and hum them as you go about your daily regimen of household tasks like making dinner, washing dishes, folding laundry, and taking out the trash. We hear these tunes more often than we hear radio-play of our favorite artists' songs. If you find yourself humming along and breaking into dance, hey, don't sweat it, you're not alone. Those songs were produced to burrow deep inside your subconscious and hang around for a while as you mull over whether you may be experiencing a symptom or side effect. Cue that catchy jingle to a once forgotten song. *Sorry, Taylor Swift!* The pharma industry has this category of music cornered. And the Grammy goes to...

Today, We are Creator and Consumer, *But to What End?*

Consider, for a moment, humanity's ancient peoples. Imagine how silly and humorous it would be to go back thousands of years and witness ancient people flip rapidly from one parchment page to the next, rolling their eyes in disdain or disgust, as they scanned the scrolls for the latest philosophy, technology, or trend. Now, of course, I'm using a play on words to draw distinction between our precious ancient writings and our modern version of scrolling on social media as an allegory.

Ancient scrolls had a deeper purpose. They were used for recording history, sharing and preserving literature, and for marking ceremonial and religious purposes. In this way, scrolls were, and continue to be, held sacred. Imagine thousands of years from now, a future generation accesses a long-forgotten data cloud, only to witness and interpret our society through the lens of TikTok reels and memes? With proper context, at least they may see that we had a sense of humor. It will be interesting to see how long our current fascination and temptation of scrolling will stand up in our current sands of time, before we rediscover our intergenerational desire for preserving ourselves and the past.

How we capture, share, celebrate, archive, and narrate life today can feel manufactured and transactional. Certainly, there is a method to our madness, and for most people, social media is but a tool for self-expression, exploration, entertainment, and communication. But as our data gets stored and backed up onto modern-day scrolls (i.e., "the cloud"), we must wonder whether we are advancing humanity with dignity and wisdom or simply streaming our innate wants and desires for attention. TikTok, Twitter, Instagram, Facebook, WhatsApp, and other social media tools are not necessarily the problem. What gets posted, shared, liked, criticized, and celebrated on social media reflects [some of] society's underlying needs, wants, desires, and values. Social media is a mirror to our existing human condition and spirit, in all its dysfunction, divisiveness, glory, and angst.

Marketers, social media influencers, even some newscasters, politicians, and business leaders understand this and want us to stay tuned in, and on point, with their message. The short-cycle attention span has led us to a culture of jumping from a breaking news announcement to the latest dance fad, to a political rebuttal, and back to another breaking news tidbit within minutes, if not seconds. As multimedia domains compete for our eyeballs and ears, our attention span wanes, homing in on the more shocking, fantastic, or unbelievable to maintain our thirst for egoistic gratification. Yet within our psyche and neural operating system, we yearn for exercising deeper personal connection and critical thinking. It's in our nature to want to experience life through our senses, in the moment, as well as through our innate ability to create, deconstruct, evaluate, and reason. Exercising all elements of our human intellect enriches our understanding and fulfillment of life.

Some might say that our virtual pursuit of happiness is fraught with clear and present issues, including social, economic, energy, and environmental impacts. Behind every scroll, swipe, and send; behind every clip, reel, and post; behind every like, love, and celebrate; behind every notification and nagging pharma song stuck in our head; there is one of the largest energy and industrial complexes in human history, standing up the "scrolls of our time." Our data-driven, hyper-digital, scrolling society is enabled by the cloud, enormous datacenters, and high-performance computing centers that provide the fuel that feeds our frenzied and fanatical foray into the scrolling abyss.

On a consumptive level, we may not see the direct correlation between scrolling, swiping, and sending and the enormous draw of power that is required to ensure new content loads every millisecond. In aggregate, however, billions of emojis and likes begin to quickly add up, as most major social media platforms consume more energy on a monthly basis than what some countries consume annually. In fact, Information and Communication Technologies (ICT) account for 6 to 10% of global electricity consumption, or 4% of our greenhouse gas emissions.[81, 82][ii][iii]

A Shot Across the Bow at "Righteous Pontification"

"I stand corrected." Have you ever "stood corrected?" If I were to post a message on social media or reply to every post that caught my attention, chances are, if someone wanted to counter my point of view (POV), I would stand corrected. There is nothing inherently wrong with this. We all have certain POVs that we want to make and let the world know. But in the past few years, social media has become increasingly toxic.

Our screens are a veil protecting the illusion that we can spew and spout whatever, whenever, wherever we want about just about any topic known to humankind and then some. For a time, perhaps this provided a necessary state of release, call it a mass experiment on human therapy on the heels of COVID. But look deeper, there is something more sinister and devious to what's happening. While social media has its utility, it has also provoked a dark side of human emotion. In some cases, the people we know, our family, friends, and neighbors, have become estranged, unhinged, and unkind.

I have deliberately slowed my active social media participation over the past couple of years. It has become clear that the modality for sharing knowledge and communication has become tainted with biased code, centered more on uncompromising POVs promulgating mis- and disinformation, and operated by a tech machinery whose values and intentions are greatly obfuscated.

Many posts I continue to see now are what I call, *Righteous Pontification.* To be clear and honest, I have been a Righteous Pontificator before. It's easy to do so, and to get caught up in wanting to be seen as smart, and on the right side of an argument. The problem with this in a digital age is that the utility of most social media platforms is limited. If you can't get your point made in 240 characters or less, then, well, don't even try. The modality magnifies the madness — a self-centered communication marked by cute selfies, memes, and videos that celebrate short, curt, incomplete rants that don't lead to any real resolve. Before you judge that I'm completely cynical, let me disclose that I certainly scroll, get lost within the content rabbit hole, and laugh out loud at videos, images, and the happenstances of others' lives.

Amid our distractions, it's easy to get caught up in pointless scrolling, or worse, seeking out the subtexts by which we can extend our pontifications so that we can feed the ego, get that hit of dopamine, and momentarily feel relevant. I don't see a comment thread on social media as communication, so much as it is righteous pontification. Sure, there is language and the perception of dialog, but isn't most of this really ego jousting, seeking those "gotcha" moments among friends and peers? Even if that doesn't sound like your personality, do you at least think about the refrain you would post, if the ego allowed you to do so? I've found all of this to be quite exhausting and pointless, leading to our own distorted sense of reality and relationships.

Discovering Deeper Truths About Ourselves, and What's Sacred During Our Sands of Time

We all know the phrase, "life is short," and have probably said it many times. We also know, as we age, that time is fleeting, passing through our hands like loose sand at the beach. The more we try to grasp for control, the quicker time can pass us by. Perhaps I'm crossing over into the righteous pontification realm here, but our time does not have to be so elusive. We may not control time, but we do have control over the quality of the time that we do have available to us. We can choose to aimlessly scroll, or we can invest ourselves in something more meaningful that brings us greater joy, and serves the people and world around us. The choice is ours from a consumer perspective. From a creator perspective, we also have a choice. We can choose to leverage technology and the platforms we have available to us to foster more meaningful connections and practical and sensible content.

We need to stop accepting the notion that people's attention span is just a few seconds or less. In reality, the most compelling, challenging, courageous, and caring questions that should (and need to) be asked of ourselves and society will require a level of cerebral immersion that most people don't [necessarily] want to exert personal energy toward.

The more we push tough questions that require deep cognitive immersion away, the further we distance ourselves (individually and collectively) from seeking truth, garnering wisdom, and achieving a deeper sense of understanding for life. Sure, it's much easier and a lot more entertaining to focus our limited time and attention span on the fleeting and fickle ideas and messages that stream onto our screens each day. But if we only accept and settle for what entertains us, we minimize our potential to elevate our intellect to prosper, let alone to be enlightened.

Points on Pragmatism

- *Scrolling is not inherently bad. It can certainly serve as a form and function as entertainment, as helpful content and information, and as a way to connect and communicate.* ***It pays to be self-aware of mindless scrolling, and the toll it takes on your mental health, not to mention the drain on time it can be.***
- ***We understand that social media is not a crystal ball that can, consistently and accurately, foretell the future or provide us with the needed intuition and wisdom to navigate change.*** *Those visceral reactions are tools that we already have, embedded in our DNA. They stem from our guts and our minds, guiding our daily behaviors and decisions. Social media feeds our programming in a ritualistic sense, whereas our innate wisdom guides a deeper sense of purpose, meaning, and survival.*
- *As consumers, our digital footprint has a material economic, social, energy, and environmental impact on our individual lives and greater society. Although our individual scrolling contribution to CO2 emissions may seem insignificant, the reality is that billions of views of Taylor Swift's Eras Tour videos add up to a significant amount of power required to store and access all of those fan favorite moments on demand.* ***We live in a frenzied, on-demand digital culture that***

loves to consume data, and subsequently, energy and natural resources. *Remember that data that is created, stored, shared, curated, etc., equates to power demand. In the U.S., 60% of the electricity generation continues to be derived from fossil fuels — coal, natural gas, petroleum, and other gases. About 19% of the electricity is nuclear-generated and about 21% is generated from renewable feedstocks.*

- *Although our modern consumer culture has perfected the art of attention-grabbing marketing, advertisement, and influencer techniques — there are deeper truths, thought-provoking issues and challenges, and human need for identity, belonging, purpose, and preservation — which require humanity to invest our time wisely into more nuanced human-to-human dialog, community discussion, and development.* ***Will (and can) the egoism and perception of individuality baked into our digital culture give rise to a more lasting opportunity for humans to be seen, heard, respected, and valued?***

11

POWER TO AND FOR THE PEOPLE: A PLAYBOOK FOR PRINCIPLED ENERGY THAT POWERS PROSPERITY

Everything humanity needs to live in harmony with each other and the planet, and to be sustainable in every sense of the word, exists today. I love technology; however, we do not require a tech-fix or next-generation anything to live more sustainably. While I teach students and coach companies and mentor entrepreneurs on sustainable enterprise and am an advocate for sustainable consumption, the truth is that we cannot exclusively consume our way to a more sustainable planet.

In recent years, greenhouse gas (GHG) emissions and in particular, carbon emissions, have led social media, policy, and political agendas. In fact, the International Energy Agency (IEA) estimated that USD 1.7 trillion[83] was invested in clean energy in 2023, including renewable power, nuclear, storage, low-emission fuels, efficiency improvements, and end-use renewables and electrification. Clearly, our focus on GHG abatement and divestment of decarbonization solutions is essential to reducing climate risks, but we must not marginalize the additional high-risk economic, social, and environmental concerns that also place significant challenges before people and planet.

The majority of my professional "day job" is directed toward advising government and industry on how they can accelerate their strategic shift toward sustainability in an equitable, efficient, and high-impact way. On August 16, 2022, President Biden signed the Inflation Reduction Act (IRA) into law, putting a capstone on the most significant piece of legislation that Congress had ever taken toward clean energy and climate change. Since that time, there has been a groundswell of investment and motivated investors, policymakers, corporations, community organizations, and other stakeholders who have been swooning at the sheer size and scale of the IRA's treasure trove of incentives.

In 2022, the U.S. electric power industry accounted for approximately 33% of total energy-related carbon dioxide (CO2) emissions.[84] The electric power industry is essential to the U.S. and global economy. Critical sectors, including healthcare, public safety and security, financial services, manufacturing, government services, transportation, datacenters, and telecommunications, are reliant on safe, reliable, and affordable electricity. Without adequate power, our economy and livelihoods would come to a halt. The electric power industry is an enormous source of GHG and carbon emissions, however, it also designs and operates the major arteries that enable decarbonized and sustainable pathways, including renewable energy development, transportation electrification, battery energy storage, and advanced sustainable manufacturing. The electric power industry sits front and center as integral to the U.S. and global shift toward a sustainable economy. As a metaphor, the electric power sector is the beating heart of a more sustainable future. However, the "health of the heart" is under duress, brought on by an aging infrastructure, rigid and outdated business models, and a spike in renewable generation projects and customer demand accelerated by the federal stimulus, enterprising investors, and state-led policy and legislative mandates.

In April of 2024, the U.S. Department of Energy's Lawrence

Berkeley National Laboratory estimated[85] that 2,600 gigawatts of new power generation and storage capacity were actively seeking interconnection with the grid, representing an eight-fold increase from a decade ago. Converging drivers, including investment, incentives, and policy, stimulated demand for clean and renewable sources of energy. The interconnection of new energy assets requires a process to evaluate how any new generation, energy storage, digital technologies, and other assets will interfere with requirements for safety, security, reliability, power quality, peak load management, power pricing, and other critical power grid operating parameters.

The flood of interconnection applications had left many regional transmission operators (RTOs) and independent system operators (ISOs) overwhelmed. RTOs and ISOs provide market facilitation for the electric power industry, ensuring grid operations are maintained reliably, efficiently, and cost-effectively. They manage energy transactions and facilitate the energy operations within broad multi-state or multi-province regions (RTOs) and localized state jurisdictions (ISOs). Before any new energy generation asset is connected to a power grid, it must file an interconnection request with an RTO or ISO. The RTOs and ISOs have had a growing backlog of interconnection applications due to limited resources and lack of employees to evaluate the huge intake of renewable and battery energy storage requests for interconnection. In April of 2024, the U.S. DOE released a roadmap[86] that sought to "outline solutions to speed up the interconnection of clean energy onto the nation's transmission grid and clear the existing backlog of solar, wind, and battery projects seeking to be built.

The Transmission Interconnection Roadmap, developed by DOE's Interconnection Innovation e-Xchange (i2X), serves as a guide for transmission providers, interconnection customers, state agencies, federal regulators, transmission owners, load serving entities (LSEs), equipment manufacturers, consumer advocates, equity and energy justice communities, advocacy groups, consultants, and the research community, which includes DOE. The roadmap sets aggressive success targets for interconnection improvement by 2030 and outlines tools that will improve the process for connecting more clean energy projects to a reliable grid, while helping achieve the Biden-Harris Administration's goal of 100% clean electricity by 2035. Time will tell if the DOE's i2X initiative and roadmap is successful in expediting the interconnection of clean and renewable energy assets on the U.S. power grid. Undoubtedly, initiatives like this need to intervene when markets display barriers or inefficiencies that inhibit their ability to effectively serve their purpose.

The electric power sector represents the epitome of classic sustainable transition challenges. Consider the following:

- The uptick in demand for decarbonized, democratized, digitized, and dignified energy solutions stimulated by investors, incentives, and policy poses a high risk to the nation's power grid infrastructure and operations.
- An increase in "behind-the-meter" distributed energy resources (DERs) and distributed energy networks has dramatically increased the number of grid-connected devices and digital end-points (and subsequently, entry-points which elevate cybersecurity and cyber-physical security concerns).
- The intake of clean and renewable energy generation interconnection applications is at a standstill. In the interim, the grid continues to have pent-up demand for new generation and transmission and distribution upgrades.
- Some regions of the U.S. are struggling with a total reduction in power generation assets as older and inefficient fossil and nuclear facilities are being taken out of service, as their license to operate was not granted under new policy frameworks and regulatory standards directed toward decarbonization goals.
- Meanwhile, the shift to decarbonized transportation, buildings, and behind-the-meter energy solutions has stimulated new demand for electricity. Electric vehicles, demand for transitional fuels like green hydrogen, net-zero buildings, and a host of other technologies and applications are calling for more renewable electricity today and forecasting a growth in future demand.

The situation for many economic growth regions of the U.S. is one where demand for clean, reliable, and high-quality electric power is increasing, and the availability of renewable grid power is limited. The advancement of a digital economy combined with commercial and consumer use of artificial intelligence (AI) has stimulated demand for datacenters, and subsequently, more electricity. In addition, favorable policy and government incentives have spurred consumer-driven demand for electric vehicles (EVs), lower-carbon-emitting building electrification measures like heat pumps, and other clean energy technologies that have further created demand for clean electricity and energy. Rising demand for cleaner electricity has been limited by the availability of renewable energy generation that has been built and integrated with the power grid, and the longstanding "power of physics" challenge associated with the intermittency of renewable resources, namely wind and solar.

Distributed energy solutions (DES) and battery energy storage solutions (BESS) have stepped up and into the renewables intermittency challenge to tackle this issue head-on. However, the large-scale siting and interconnection of BESS with renewable systems remains limited, constrained by a host of market-based, community-development, and regulatory hurdles including a new era of community groups that don't want BESS or renewable generation in their backyard (i.e., stakeholder groups that oppose unwanted development in or near their neighborhoods through campaigns, including Not in My Backyard, NIMBY; and Build Absolutely Nothing Anywhere Near Anything, BANANA). While technical challenges like intermittent power, energy storage, and power quality can be overcome, the solutions add additional community buy-in and regulatory approval, financial costs, and technical requirements. Currently, our society is experiencing a triple threat to the U.S. power grid with critical risks including (1) an overly constrained grid that has high power demand and inadequate dispatchable generation resources to meet that demand; (2) a generally outdated transmission and distribution system that, in its current state, is not prepared for onboarding new renewable generation without significant investment to support power intermittency; and (3) a grid that is being challenged by new business paradigms including the advance of a digital society that includes more opportunities for cyber-intrusion and cyber-physical security threats. The existing input-output energy infrastructure that once served our needs is no longer optimized to serve today's digital economy and newfound electric demand. It's outdated, inefficient, and does not currently integrate the age-old wisdom and principles we have for attaining a more sustainable system and future.

Our power grid and system are constrained, but it is also undergoing a massive once-in-a-generation shift to better serve the evolving needs of business and society. The "traditional electric utility model" was constructed based upon a command-and-control centralized market that provided a linear, one-directional flow of electricity and value from generators to consumers (see figure below).

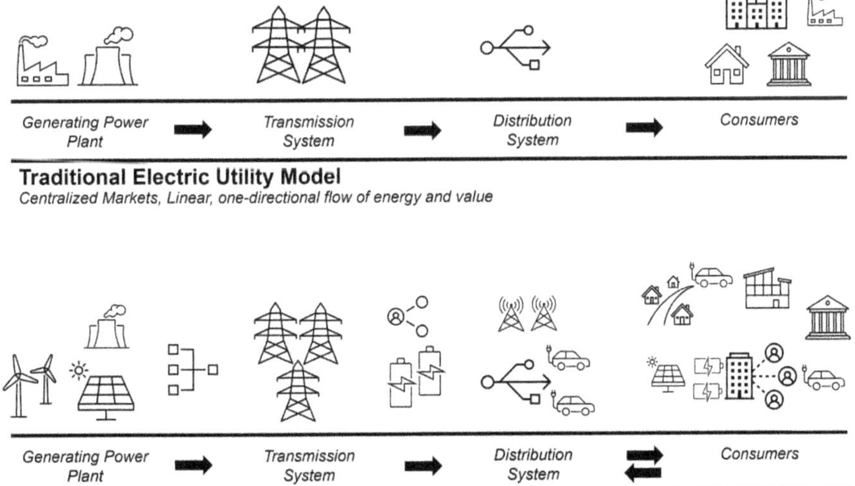

In contrast, the evolving "transitional distributed electric utility model" is working to modify the existing market structure of the

traditional electric utility model toward a more decentralized, non-linear energy market where decarbonized, digitized, and democratized energy solutions reside in the hands of producers and consumers, giving rise to a multi-directional energy flow and multiple value streams (opportunities for creating and transacting energy and economic benefits). If we utilize principles of pragmatism in the design, engineering, construction, upgrade, operations and maintenance, and long-term planning of the electric utility and power grid, we can enable a more resilient and sustainable decentralized energy marketplace.

A Clear Focus on Nuclear Energy

One of the most iconic NIMBYs (not in my backyard) in the energy sector is nuclear power generation. As a technology, nuclear has an enigmatic and troubled past. During the final stages of World War II, the U.S. atomic raids on the Japanese cities of Hiroshima (August 6, 1945) and Nagasaki (August 9, 1945) marked the first and only time nuclear weapons were used by any nation during a conflict. The destruction was unfathomable. Approximately 105,000 Japanese and Koreans were killed by the two bombs. The nuclear bombings also led to the end of the Pacific War and subsequently the conclusion of World War II. However, following the Hiroshima and Nagasaki nuclear bombings, the research and proliferation of nuclear technology and armaments rose, particularly between the United States and Soviet Union. Other nations, including India and Israel, also initiated nuclear weapon development. The development of nuclear weapons was thought to serve to demonstrate scientific and innovative prowess and as a militaristic tool for deterrence. Following the very first nuclear detonation on July 16, 1945, what was known as the "Trinity nuclear test," J. Robert Oppenheimer, the Manhattan Project lead manager, stated[87],

> *"We knew the world would not be the same. A few people laughed, a few people cried, and most people were silent. I remembered the line from the Hindu scripture the Bhagavad Gita. Vishnu is trying to persuade the prince that he should do his duty and to impress him takes*

on his multiarmed form and says, "Now, I am become Death, the destroyer of worlds." I suppose we all thought that one way or another."

—J. Robert Oppenheimer, The Decision To Drop The Bomb

Death and destruction, some of the original design intentions, also became the preeminent face of nuclear technology for the past eight years. This is unfortunate because, as an energy generation technology, nuclear energy offers humanity an incredible opportunity for enhancing our quality of life while also respecting the planet. As a broad category of technology, nuclear has had a profound influence on our lives, catalyzing human innovations including powering submarines, offering breakthroughs in medical imaging and diagnostics, agriculture, space exploration, energy generation, and even for criminal investigation[88]. From an energy generation perspective, it's interesting to point out that:

- 20% of the electricity generated in the United States is derived from nuclear, representing the single largest source of clean energy in the U.S.
- Nuclear power facilities operate in 32 countries worldwide, generating approximately a tenth of the world's electricity.
- Nuclear generation represents nearly two-thirds of the electricity generated in France and Slovakia. Nuclear generation represents nearly one-third or more of the electricity generation in the countries of Belgium, Bulgaria, Czech Republic, Finland, Hungary, Slovenia, South Korea, Switzerland, and Ukraine[89].
- Nuclear technology continues to be researched and studied, bringing forth new innovations and applications, such as advanced Small Modular Reactors (SMRs). Advanced SMRs[90] offer the potential to size nuclear generation from tens of megawatts to hundreds of megawatts (MWs) and to be co-located near electricity demand for uses including desalination

of water, industrial utilization, high-density computing and datacenters, and other energy demand for processing heat and reliable electricity. SMRs offer advantages over their larger and centralized nuclear reactor sites in that SMRs have a much smaller physical footprint and lower cost of capital investment. This enables SMRs to be in areas that large nuclear facilities could not be sited. Further, SMRs have unique safeguards for safety and security, further making them attractive compared to the risks and concerns often cited by local citizens.

- Nuclear fusion energy, much different than the fission reaction nuclear process, is one of the most environmentally friendly sources of energy — if we can consistently ensure a safe reaction, at-scale, and in an affordable manner. The allure of fusion energy is that the process emits no CO2 or other atmospheric emissions that contribute to global warming. The fusion process uses two sources of fuel: hydrogen and lithium. Although these fuels have their own challenges regarding availability and production capacity, the fuel sources are much more stable than the isotopes required for fission. According to the International Atomic Energy Agency (IAEA), fusion devices produce approximately ten MWs of fusion power. The Fusion Industry Association[91] estimates that more than $6 billion USD has been invested in fusion energy development. Billionaires, including Jeff Bezos, Richard Branson, and Bill Gates, have invested in leading-edge fusion development companies, notably Commonwealth Fusion Systems (CFS) and General Fusion. The investment in fusion energy has stimulated strong interest in the technology's development and future. Proof-of-concept fusion reactors are being designed and developed today; however, the scaled development of fusion energy is anticipated to take twenty to fifty years or more, based upon current projections.

Clearly, advanced nuclear energy technology and power generation are going through a renaissance. The potential for SMRs as a transitional technology, supporting the decarbonization of the power sector and grid, is significant. SMRs provide the option to co-locate power generation near high electric demand requirements serving advanced manufacturing and datacenters. SMRs can also complement the transition from fossil-based generation to clean generation, including providing localized power generation baseload to supplement the interconnection of intermittent renewable resources such as wind and solar.

In early summer 2022, I had the opportunity to participate in a cross-sector industry discussion on[92], *"Reliable Decarbonization in the Northeast — Dialogues, Policies and Innovation,"* where I had the opportunity to hear from Mr. Joseph Dominguez, President and CEO of Constellation Energy, who was a feature keynote speaker at the event[93]. Mr. Dominguez oversees Constellation's clean energy generation fleet, which includes nuclear, wind, solar, hydroelectric, and natural gas facilities located in 19 states. Further, under Mr. Dominguez's leadership, Constellation generates competitive energy for over two million residential, public sector, and business customers nationwide, representing more than three-fourths of the Fortune 100[94].

At the June 2022 discussion on reliable decarbonization, Mr. Dominguez began his remarks with a brief orientation of Constellation by highlighting the following key company statistics:

- Constellation employs 13,000 workers;
- The company has a power generation capacity of more than 32,400 MWs;
- The company services more than 20 million homes and businesses with clean energy;
- The company has revenues of more than $17 billion and total assets of $49 billion;
- Constellation comprises the nation's lowest-carbon power generation fleet among large power producers; and,
- The company produces over 10% of the nation's clean energy, aiding America's transition to a clean, sustainable future.

With 21 nuclear reactors, Constellation operates the largest fleet of nuclear plants in the U.S. This fact is important, as it provides context to Mr. Dominguez's position and point of view on how the U.S. can explore decarbonization pathways that are pragmatic, that is, they enable growth and innovation while continuing to deliver affordable, reliable, and low carbon energy. Mr. Dominguez laid out his vision for a decarbonized future, one that leverages and optimizes Constellation's clean energy resources, but one that also acknowledges the potential for existing nuclear energy to help the nation transition to cleaner fuels and more sustainable production and consumption. Mr. Dominguez remarked that nuclear facilities, although highly efficient, have an opportunity to leverage their waste heat toward thermal energy networks, further creating efficiency gains while providing heat or steam for localized needs. He also pointed out that some of America's nuclear facility fleet is aging and coming up against its decommissioning cycle. He suggested that a review of the nuclear fleet based upon age, performance, and decommissioning timelines could provide an opportunity to explore the development of SMRs and/or other clean energy generation to supplement those generators proactively. Further, in doing so, the older nuclear facilities could be explored for "taking off the grid" and utilized exclusively as a transitional power source for green hydrogen production. In that scenario, the power from older nuclear plants could be directed toward electrolysis, a carbon-free process that uses clean electricity to split water into hydrogen and oxygen.

As I listened to Mr. Dominguez's keynote, my ears perked up. I scanned the room of energy industrial sector executives, searching for anyone who also shared the enthusiasm I felt. What Mr. Dominquez laid out was the promise of a decarbonized future brought forth by principles of pragmatism. Highly astute to the drivers and constraints of his industry, Mr. Dominguez acknowledged and conveyed the regulatory, market, and political challenges that stood before the energy sector's transition to lower carbon fuels and electricity production. In my words, he presented the need for stakeholders to go back to school on the "physics of power," referencing that many policy goals and objectives for lower carbon energy resources do not pencil out when you explore the requirements for energy density, reliability, affordability, and the existing transmission and distribution constraints on the power grid.

The U.S. energy sector and electric utility market, specifically, are vastly regulated. The market structure has served the design and advancement of the industry for over a century. During this time, there have been significant advancements in how electricity is used by industry and society. Advancements in technology have also introduced new ways in which we can optimize our production and use of electricity. The traditional business models of the energy and electric utility sectors, constructed on regulatory and market structures of the past, cannot remain static if we are to address the shifting needs of power users and decarbonization goals of policymakers. Mr. Dominguez's talk essentially spoke to this point: that industry is prepared to change, but to transition to a clean energy future, the underlying regulatory and market structures that incentivize the traditional energy model also need to evolve. Applying principles of planet pragmatism, we must rethink not only our technology and business models, but we must also evaluate and update the market and regulatory structures that govern our energy markets to ensure the fair and competitive advancement of a clean energy future.

The inclusion of nuclear generation as a critical resource to support a sustainable energy future is proactively being pursued in other parts of the world as well. Over the past two years, I've had the pleasure of getting to know a nuclear technology enthusiast and advocate, Princess (Princy) Mthombeni of South Africa. Princy is the Founder of Africa4Nuclear, an advocacy group that promotes nuclear energy as a key contributor to Africa's Agenda for Sustainable Development. In getting to know Princy, I've invited her on a few occasions to speak to my undergraduate and graduate business school students at Syracuse University, offering her perspective on the challenges and opportunities tied to sustainable development in South Africa, including the role of nuclear energy. In the spring of 2024, Princy spoke before my graduate class, "Managing Sustainability," with the following details and perspective on Africa:

- 17% of the world's population resides on the continent of Africa.
- Africa has the lowest amount of electric power generation. Only 4% of the world's electric power is generated in Africa.
- An estimated 600 million people in Africa don't have access to electricity, and approximately 900 million people lack access to clean cooking.
- To serve the needs of a growing population, Africa's electricity demand is projected to triple by 2040.
- The United Nations predicts that Africa's population will double to reach nearly 2.5 billion people by 2050.

Discussing challenges faced by South Africa and throughout Africa, Princy emphasized the "triple threats" of poverty, inequality, and unemployment as the key areas that she and her organization focus time and resources on in support of sustainable development in the country and throughout the continent. Princy pointed out that Africa's transportation and agriculture industries are heavily dependent upon fossil fuels. Princy provided a compelling discussion on the pervasiveness of poverty, pointing to the accessibility of affordable, clean, reliable, safe, and secure energy as one of the root causes. For developing regions of the world, including Africa, energy development offers an opportunity to stimulate job creation and economic development. In doing so, local communities have a pathway for economic and energy security, two essential pillars that can dramatically enhance quality of life by provisioning the energy and financial resources required for other public needs, such as healthcare, education, and the advancement of other social and economic opportunities. Drawing upon her deep background with nuclear energy (see Profiles in Pragmatism, Princess Mthombeni), Princy discussed the merits of nuclear energy, including the benefits tied to education and training, workforce development, and the potential to generate clean, affordable, reliable, and safe power.

Profiles in Pragmatism

Princess Mthombeni

Thought Leader | Founder, Africa4Nuclear

In today's world, an "influencer" is someone who has a strong social media presence and following, and who influences the actions, namely the buying habits, of others. Princess (Princy) Mthombeni of South Africa is much more than an influencer by any measure and contemporary definition. Princy is a leader in the purest sense. An educator, youth advocate, community builder, thought leader, communications specialist, and humanitarian — Princy Mthombeni's leadership in South Africa and throughout the world has provided a positive influence toward fostering a "sustainability generation." Princy is the epitome of leadership in action. Princy is proactively shaping the prosperity of our generation and future generations. Like many global influencers, Princy smartly leverages social media and communications technologies. But it is her dedication to create local and global awareness and action on critical sustainability concerns such as health, education, and economic progress that runs far deeper than the transactional likes sought after by most social media personalities, which sets her apart. In this era of social and environmental change, Princy's leadership is a beacon of hope, transformation, and renewal.

Princess Mthombeni, widely known as Princy, is a multi-award-winning communications strategist, an alumna of the Graduate School of Marketing, and a passionate advocate for nuclear technology. A World Nuclear University fellow from eNanda, KwaZulu-Natal, South Africa, she is globally recognized for driving high-impact nuclear communication strategies that shape policy and public perception.

As the founder of Africa4Nuclear, Princy champions nuclear energy as a cornerstone for Africa's sustainable development. With over a decade in the nuclear sector, she has established herself as a leading voice in energy discourse across the continent.

Princy's expertise has earned her speaking engagements at high-profile global forums, where she shares insights on the role of nuclear energy in shaping Africa's future. Her thought leadership has been featured in major international media outlets, amplifying her impact in the global energy space.

Her contributions have been recognized with numerous accolades, including the Women in Nuclear Global Excellence Award and the Presidential Award from the Black Business Council. She has also served on the boards of Women in Nuclear South Africa (WiNSA) and the African Young Generation in Nuclear (AYGN), continuing to advocate for the peaceful and impactful use of nuclear science and technology.

Connect with Princy:

Twitter/Instagram/TikTok: @princymthombeni

LinkedIn: Princess Mthombeni

Waste Not, Want Not: Just Remember, The Laws of Thermodynamics Cannot Bend

When it comes to humanity's need for and use of energy, we must get out of our own way, already! When we embrace a mindset of plant pragmatism, we place a more complete set of values on the resources we consume and don't expend precious energy unnecessarily. This sentiment holds true for the energy that powers our society and the energy that fuels our body and soul. Where, when, how, and why we consume energy (physical and emotional) says a lot about our priorities, intentions, and the resulting impact we see (i.e., environmental pollution, social challenges, personal fear or anxiety, etc.). Yet resource consumption alone is not the full answer for addressing rising global energy demand and the associated social, economic, and environmental impacts. We must also think about how we can optimize the productive impact of all forms of energy that are already "in-process" and available to us.

We are one quarter into the twenty-first century, and yet we remain largely arrogant, inefficient, and destructive in our use of energy and natural resources. For the past fifty years, the U.S. Department of

Energy's Lawrence Livermore National Laboratory (LLNL) has been developing energy flow charts to "help decision makers to visualize the complex inter-relationships involved in managing our nation's resources[95]." LLNL's flow chart[96], "Estimated U.S. Energy Consumption in 2023," is shown below. LLNL also produces interesting and useful flow charts that illustrate carbon and water flows.

The chart below visualizes the U.S.'s total estimated domestic energy consumption, which was estimated by LLNL to be 93.6 Quads in 2023. Examine the chart, and you can see that the U.S. energy mix is diverse, representing solar, nuclear, hydro, wind, geothermal, natural gas, coal, biomass, and petroleum resources. The LLNL energy flow chart also presents which sources contribute to electricity generation, and further, the subsequent flow of energy resources to downstream energy sectors (i.e., residential, commercial, industrial, and transportation). For years, as an energy analyst, I studied and used this chart and subsequent state-level energy flow charts to characterize total consumption.

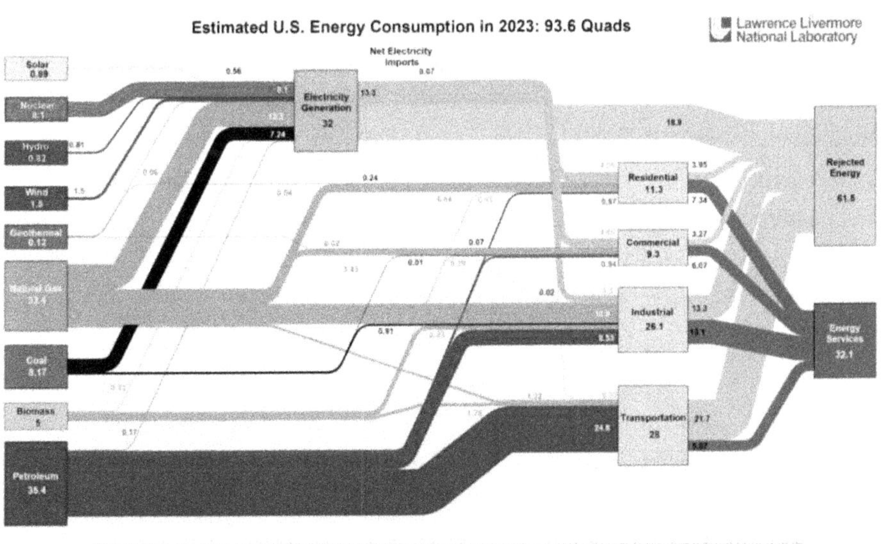

Image courtesy of Lawrence Livermore National Laboratory

I have frequently used this chart when teaching undergraduate and graduate courses in sustainable enterprise. Take another close look at the chart. Does anything else stand out for you? If you look beyond the primary energy sources and flows, you can see that 32.1 Quads of energy go into energy services, which represent the conversion of energy for "social good," such as lighting, transportation, maintaining our comfort through building heating ventilation and air conditioning (HVAC) systems. Look further at the chart, and a whopping 61.5 Quads of the energy flows lead to rejected energy. LLNL's rejected energy is characterized as the portion of energy that goes into a process that is dispensed as waste heat into the environment. Most of the rejected energy is waste heat from the burning of fossil fuels. A smaller portion of rejected energy is attributable to transmission losses for electricity.

As a sustainability educator and practitioner with a focus on clean energy, energy efficiency, and decarbonization, I refer to this chart in support of planet pragmatism. For my career, a great deal of regulatory oversight, proactive policy focus, sustainable investment, and technology development has been on diversifying the primary energy sources. State-driven Renewable Portfolio Standards (RPS) and Clean Energy Laws have been established to stimulate demand for clean and renewable energy generation resources. The success of policy and market incentive-based programs cannot be understated. In fact, according to the IEA[97] about 86% of newly commissioned renewable energy capacity (in 2022) had lower costs than fossil fuel-fired electricity. The growth of renewable generation and cleaner energy fuels has grown significantly over the past thirty years. Yet during this time, so too has overall global demand for energy. Between 2014 and 2023, total U.S. energy consumption fluctuated between 88 to over 98 Quads. Rejected energy also fluctuated over this period, hovering between 58 to 64 Quads. However, the proportion of rejected energy compared to energy services has remained relatively constant. This tells us that although society has diversified its primary energy portfolio to include clean and renewable energy, there remains a significant opportunity to reduce greenhouse gas emissions (GHGs) through energy efficiency, fuel switching, and capturing heat losses.

Years ago, when working in applied research and technology development at Rochester Institute of Technology (RIT), a colleague told me that there is no such thing as waste heat. My colleague, a senior materials research scientist and engineer, educated me to the fact that inefficient processes and products will emit "waste" as heat, emissions, or other byproducts associated with incomplete conversion of energy (i.e., chemical, electrical, thermal, mechanical, radiant (light), sound, or nuclear). Acknowledging my colleague was more educated than I in certain sciences and engineering, I still chose to challenge him, albeit unsuccessfully. Attaining 100% efficient transfer of energy goes against the Laws of Thermodynamics[98] and is not (to our current level of understanding), possible. My colleague aptly pointed out, however, that although energy transfer may not achieve parity, the ancillary byproducts of inefficient products and processes should be thought of as value streams as opposed to waste. By thinking of heat as a commodity, not as a wasteful byproduct, we can think more level-headed about best options for heat recovery, transfer, utilization, and conversion — thereby drawing as much social utility and economic value from this resource, as well as the original primary resource that generated the heat in the first place, as possible.

Now, when I study the LLNL energy (water and carbon) flow charts, I do so with a more critical and discerning eye. Much like perspectives on climate adaptation vs. mitigation — the planet pragmatist must evaluate energy opportunities holistically. Planet pragmatism beckons us to challenge convention and reconcile contradiction. It is entirely possible that too often a myopic view of the future clouds reasonable strategies that could yield beneficial impact. Subsequently, our journey toward a sustainable and prosperous future may require us to accept truths, redefine how we communicate value, and work to have common sense for the common good solutions. At the fulcrum of every economic, environmental, and societal challenge we face is human interaction, engagement, and interference.

The world does not necessarily have an energy crisis, rather, we have a misunderstanding, misallocation, misalignment, and mismanagement of resource crisis. By understanding the sustainability truths that lie visibly in front of our generation, we can and shall redefine resource utilization in terms of planet prosperity. For example, the decarbonization of our global economy can happen through the greater diversification of primary energy sources, including clean and renewable energy. Yet, at the same time, if we do not focus our attention on simultaneously and drastically minimizing the rejected energy that is embodied within our existing products and energy infrastructure, developed countries will only continue to think of primary energy sources as binary inputs that are to be consumed, and their corresponding byproducts to be classified as waste. The pursuit of circularity and more holistic systems thinking represents tools for pursuing planet pragmatism. We can learn a great deal by focusing not only on energy diversification, but just as much on how we optimize for the "in-process" energy sources and systems that currently power and fuel our homes and our global economy. *Going forward, we need to use precious resources, especially energy, with precision so that it may optimize its transference into economic and social good, serve as a catalyst for positive social and economic change, and advance prosperity through principles of planet pragmatism, which benefit all of humanity.*

"Feed Me, Seymour!" Is Generative AI the *Little Shop of Horrors* of our Generation?

Emergent generative AI models, including the popular Generative Pre-trained Transformer, or GPT, are exciting tools, offering advanced capabilities for information analysis and digital creation, elevating the potential for a societal renaissance enabled by AI technology. But for all the allure, fascination, hype, and benefits promised by artificial intelligence (AI), there is a dark side to the burgeoning technology.

The algorithms underlying Generative AI models are rendered unproductive unless they are constantly fed. For AI models to be effective, they must grow, learn, and be continuously stimulated. It's not that different than raising a baby. To grow and develop, a baby needs to be nurtured and nourished. Babies need to eat, explore, be taught, and loved. AI has an insatiable hunger and thirst for data, energy, and even water — yes, water. Much like the masked shrew[99], generative AI has a very high metabolic rate. The masked shrew eats three times their weight a day and they can only survive a few hours without eating. AI operates 24-7 and needs to continuously feed (data, energy, natural resources); otherwise, it will atrophy and die.

AI models including OpenAI GPT-4, Meta AI, Microsoft Copilot, Google Gemini, and DeepMind, among others, have captured the imagination of a new generation of tech enthusiasts who are leveraging the immense computing power of AI engines to gain a creative and competitive edge as they improve, accelerate, and optimize their digital productivity and impact.

Next-generation tech builders have flocked to Generative AI and other AI models for benefits such as automating repetitive tasks, enhancing decision-making by identifying patterns and trends found in analyzing huge amounts of data, improving workflows, minimizing human-induced errors, boosting productivity, and reducing costs. However, although benefits abound, the AI mantra of better, faster, cheaper comes at a significant economic, environmental, and societal cost to their owners operating the AI engines, and to all of us comprising broader society.

Is Generative AI a Little Shop of Horrors for this Generation?

Generative AI is akin to Audrey II from the classic cult movie, *Little Shop of Horrors.* In the movie, Audrey II was a pet plant, a Venus flytrap, that had a demented plan to dominate the world. To accomplish this, Audrey II needed to be fed human blood to stay healthy and grow. Audrey II befriended Seymour, a worker in a florist shop, who persuaded unsuspecting victims to visit the flailing flower shop, so that they would be devoured by the devious carnivorous plant. The famous line, "Feed me, Seymour!" delivered in a demanding voice by Audrey II, echoed in the minds of moviegoers far after they left the theatre and went about their daily lives.[100]

As Audrey II gulped and grew, the plant became a prime attraction for the florist shop, luring in more and more customers, and ultimately, victims. This continued as a self-reinforcing cycle until the movie climax, when Seymour tried to destroy the beastly plant he had helped grow into a monster. Ultimately, Seymour suffers great loss when his original love, a woman named Audrey, succumbs to a fatal wound and is consumed by Audrey II. Seymour positively changed the shop's financial fate, but at great personal loss. If you've never seen the 1960 original or the 1986 remake of Little Shop of Horrors, check it out. The movie creatively and humorously explores the evergreen themes of greed and survival, egoism and humility, capitalism and unfettered growth.

There is no such thing as a stupid question, but there are costly ones.

You've likely heard or even said the phrase, 'there is no such thing as a stupid question.[101]" We are taught and coached by teachers, parents, grandparents, and bosses that we should not be afraid to ask questions, even if we believe that they are ridiculous. Typically, in the act of asking a question we perceive as silly, we aid others who may have the same, or similar, question. I know that as a student and a professional, I have squirmed and sunk lower in my seat, not wanting to ask an obvious question, fearing ridicule from my peers. I have also perked up when I've heard the answer to a question someone else provoked, and that I was too afraid to ask. We've all had these moments of feeling unnecessarily inferior and foolish among our peers.

The interesting thing about Generative AI (Gen AI) tools is that they remove the perception of judgment that drives our fear and fuels our hesitancy to ask questions. They provide us with a safe space, so to speak, to be vulnerable without repercussion. The point with Gen AI is to use a conversational tone as you phrase a prompt. The Gen AI then provides you with an answer. The answer may or may not be functionally and intellectually helpful, but it is delivered without judgment, thereby protecting your psychological well-being.

Gen AI is a powerful tool with capabilities and intentions that extend beyond the prompt of a simple question. That said, who hasn't asked ChatGPT or another Gen AI tool something esoteric or pertaining to existentialism, even for fun? Does ChatGPT know the meaning of life? Can ChatGPT tell me how to find love, make more money, cure a disease? Experts will tell you that ChatGPT and other Gen AI do not do well answering questions that require deep reasoning based upon the nuanced understanding of language or analysis of multiple complex ideas or topics. That said, some experts believe that AI has the potential to become more sentient in its capability to provide inference on a humanistic level. Until that time, we are stuck with the more rudimentary version of Gen AI.

While they cannot answer your most pressing existential questions, the current versions of Gen AI are incredibly powerful. With the right prompts, Gen AI tools can do amazing things that deliver quantifiable improvements in your productivity. Gen AI tools can read and write computer code, assimilate and analyze data and information, write a story or even a book, provide comparative analysis, and so much more. Gen AI is rapidly expanding its reach and capability. Each day, there are new use cases of the vast potential Gen AI can provide to business and society.

AI and DataCenters: Feed me, Seymour!

Underlying Gen AI's magic and impressive advances are its steep operating requirements, including continuous energy. Like the fictional Audrey II from Little Shop of Horrors, ChatGPT and other Gen AI tools need to be fed, all the time. While they are not asking for human blood (like Audrey II), Gen AI does need massive amounts of reliable electricity. If you were not aware, there is a datacenter boom[102] that is underway in the United States, and throughout the world.

Bain & Company, a management consulting firm, estimates that the surge in datacenter development and power consumption could require more than $2 trillion in new energy generation resources worldwide. Further, Bain & Company analysts project that U.S. energy demand could outstrip supply within a few years, requiring electric power utilities to increase power generation. Bain & Company analysts are projecting (in their high-case scenario), a 44% increase in electric load growth for datacenters between 2023 and 2028, the highest load growth among customer segments (including commercial, manufacturing, and residential segments).

Wow, that is a lot of electricity. "Feed Me, Seymour!"

In the US, swift spiking demand for electricity to power datacenters that enable Gen AI is pushing the limits of the existing power grid and available power generation resources. As a result, market-based partnerships are forming to proactively develop clean, reliable, and abundant generation resources — as well as reduce the impact of localized datacenter operations. For example, in the fall of 2024, Microsoft and Constellation Energy[103] announced a deal to recommission the Three Mile Island nuclear power facility in Pennsylvania as an uninterruptible clean power source to feed Microsoft's rapidly increasing datacenter power demands. Meta also announced[104], in late summer of 2024, a partnership with Sage Geosystems, to significantly expand the use of geothermal energy in the US. Leveraging Sage's Geopressured Geothermal System (GGS), Meta aims to have carbon-free power distributed for use within its datacenters. Meta, which has already contracted more than 12,000 MW in renewable energy projects, will work with Sage to scale their GGS technology, with the first phase of their project, development and delivery of 150 MW of geothermal energy for Meta datacenters to be operating by 2027.

Gen AI's resource consumptions are primarily driven by two essential activities: (1) training large-scale models, and (2) running inferences, or using the model to generate responses. On the AI model training side, Gen AI is starved for greater amounts of data and power. It is becoming increasingly clear that many big tech companies are leveraging their social platforms as data lakes that support some of their Gen AI model training. Our data, including the "expert" insights we get invited to share on some platforms, are helping to train AI models. Without knowing, we all have been lured into the proverbial florist shop and are a part of feeding the beast, whether we realize it or not!

According to The Washington Post[105], prompting an AI chatbot like GPT-4 to produce a 100-word email just one time requires the equivalent of one bottle of water and the electricity needed to power 14 LED light bulbs for one hour. The Washington Post reported that in its training of GPT-3, Microsoft's datacenter used 700,000 liters of water, and when Meta trained its LLaMA AI model, their datacenter operations used 22 million liters of water. Water is as important to datacenter operations as electricity. Water is used to cool datacenters, ensuring the servers and electronics do not overheat. The Washington Post estimates that if 1 out of 10 working Americans (roughly 16 million people) were to prompt GPT-4 to create the 100-word email once a week for a full year, the datacenters would use more than 435 million liters of water — an amount equal to the water consumed by all Rhode Island households for 1.5 days. Wow, that is a lot of water. ***"Feed me, Seymour!"***

Preventing Gen AI from becoming the Little Shop of Horrors

Datacenters provide the backbone to the advanced and fast computing power that Gen AI requires to learn (build its model), and for spontaneous response to our relentless prompts (inference). Our desire to leverage Gen AI to do more, faster, better, and cheaper has only increased its voracious appetite for more data, electricity, and subsequently additional operational resources, including water, that are all used to push the limits on datacenter productivity.

There are morals to every [good] story, including fictional movies like the *Little Shop of Horrors*. At this point in its evolution and adoption, we must consider that Gen AI is ostensibly and eerily similar to Audrey II. And we, like Seymour, are doting and in love [with AI], and subsequently caught up and coerced into feeding the beast we've created. This begs the questions:

- How big will the AI beast grow before it conquers and consumes us? Does it have plans to, and will it, dominate the world?
- What lessons can we learn from Seymour's flawed character that we can use to prevent Gen AI from becoming the *Little Shop of Horrors* of our generation?
- What role will Gen AI and other forms of AI serve in our pursuit of prosperity? How can we bring forth principles of Planet Pragmatism so that we can attain a greater quality of life without over-consuming planetary resources and degrading the natural world in the process?

Energy Pragmatism is Essential to Planet Prosperity

The availability of affordable, abundant, reliable, and clean energy is essential to the operation of our modern society and our ongoing pursuit of prosperity. Currently, there is no energy source that can singularly deliver all these values to humanity across the entire world. Global energy demand is met through a diversity of energy sources.

Given humanity's current state of technology adoption, the continued pursuit of a diverse mix of energy resources to address global demand is, in a word, pragmatic. Perhaps, with additional breakthroughs in materials science, engineering, and innovation in nuclear fusion, society will have an opportunity to decouple itself from fossil-based energy sources in the next fifty years. For now, however, a multi-pronged energy portfolio is necessary to provide the world with agility and adaptability through energy optionality, as it pursues economic growth, innovation, and attainment of quality-of-life indicators. Scaled-up, energy optionality delivers upon the existing energy demand of society. However, inherent in an energy optionality strategy are trade-offs. Although energy optionality is a pragmatic approach given humanity's current state of energy technology and knowledge, exercising principles of planet pragmatism will require us to continuously pursue energy alternatives that also deliver greater efficiency, less waste, and lower pollution.

The planet's ability to provide us with clean water, clean air, and non-polluted lands is foundational to our capacity to survive and our overall quality of life. A healthy and clean environment should be a human right. Humans are guardians of the planet. Having the right to use the Earth for survival and for pursuit of prosperity also means that we accept responsibility for our actions and that we have an obligation to preserve, protect, and restore the planet. In general, humans have not been serving as caretakers for the planet. Heck, we struggle to be caring for one another. The playbook to planet prosperity is not that complicated. It is quite simple and beautiful. When we realize that peace begets prosperity and that benevolence trumps all hatred, we will rediscover what we've always deeply known. Everything in the Universe is interconnected. Our human experience is a shared experience, yet it's but one dimension of the Universe's intelligence and the meta-physical and spiritual realm that we have only scratched the surface of. As sentient beings, there is great joy for humans to celebrate all life, uplift spirits when they are down, and provide the means for all people to thrive, not just survive.

The global energy paradigm must adapt and evolve if we are to minimize the adverse impacts of the current energy lifecycle (i.e., upstream resource mining, materials processing, refining, transportation and logistics, generation, distribution, storage, use, and waste management). Energy optionality enables humanity to modulate its utilization of energy sources based upon our diverse needs. For example, the high reliability and uninterruptible electric power requirements of datacenters or the surgical operating room in hospitals are vastly different than the needs of residential homes or non-essential commercial businesses.

We should seek to minimize the over-exploitation of energy resources, such as coal and oil, that carry a heavier environmental and societal cost — and we should seek to optimize those resources that optimize quality-of-life, for example, energy conservation and efficiency, thermal energy networks, and advanced nuclear technologies. Not all energy sources are created equally. Some carry a higher carbon footprint, others have higher economic costs, and others pose a risk or threat to national security. This is not to pit certain energy resources against one another. Through a prioritization method that accounts for all of the associated benefits, opportunities, and tradeoffs against our shared goals and need — energy can be further optimized to ensure business, consumers, and society reap the full value of its greatest benefits and minimize any negative externalities.

The world's energy mix has evolved to its current playing field based on a complex history of money, power, and rules. Within this, the energy mix has been shaped by the forces of supply and demand, government policy and incentives, regulatory oversight, command and control market structures, doctrines for global defense and national security, and numerous geo-political frameworks and actors. The global energy landscape represents a complex mosaic of market actors and resources. Restructuring the global energy paradigm is possible, but it will take time. Guided by principles for planet pragmatism, the world can leverage its knowledge to prioritize the safe, efficient, reliable, affordable, equitable, and sustainable production and use of energy.

Points on Pragmatism

- *Going forward, we need to use precious resources, especially energy, with precision so that it may optimize its transference into economic and social good, serve as a catalyst for positive social and economic change, and advance prosperity through principles of planet pragmatism, which benefit all of humanity.*

- *The electric power sector is the beating heart of a more sustainable future. However, the "health of the heart" is under duress, brought on by an aging infrastructure, rigid and outdated business models, and a spike in renewable generation projects and customer demand accelerated by the federal stimulus, enterprising investors, and state-led policy and legislative mandates.*
- *If we utilize principles of pragmatism in the design, engineering, construction, upgrade, operations and maintenance, and long-term planning of the electric utility and power grid, we can enable a more resilient and sustainable decentralized energy marketplace.*
- *Applying principles of planet pragmatism, we must rethink not only our technology and business models, but we must also evaluate and update the market and regulatory structures that govern our energy markets to ensure the fair and competitive advancement of a clean energy future.*
- *Advanced nuclear energy technology and power generation are going through a renaissance. The potential for small modular reactors (SMRs) to serve as a transitional technology, supporting the decarbonization of the power sector and grid, is significant. SMRs provide the option to co-locate power generation near high electric demand requirements serving advanced manufacturing and datacenters. SMRs can also complement the transition from fossil-based generation to clean generation, including providing localized power generation baseload to supplement the interconnection of intermittent renewable resources such as wind and solar.*
- *Advanced technology, including artificial intelligence (AI), presents both challenges and opportunities for humanity. The datacenters and high-density computing infrastructure that give life to AI require a continuous and uninterrupted supply of energy. As new electric demand from AI, electric vehicles, and building electrification grows, the existing power grid will experience power quality, reliability, and resilience challenges. Further, centuries-old utility business models will be challenged as clean, affordable, and pragmatic energy solutions are introduced at the grid-edge, supporting new distributed, democratized,*

decentralized, and digitized energy solutions for customers. While it seems counterintuitive and counterproductive to adopt AI in an age of so much energy inequality, when we lean into principles of planet pragmatism as a foundation for leadership and strategic resource management, we can leverage the speed and efficiency of AI to more rapidly address our most pressing energy challenges.

PART IV

THE NEW PATH TO PROSPERITY

12

PROSPERITY "ON THE LINE": WHAT FX'S THE BEAR AND FIRST JOBS CAN TEACH US ABOUT PURSUING A BETTER FUTURE THROUGH PLANET PRAGMATISM

A Prelude to a "Bear" of a Metaphor

Truth be told, sometimes I cringe when I see others [or myself] trying to draw insight and wisdom from pop culture. Sometimes photos, memes, song lyrics, streaming shows, and celebrity quotes should simply stand on their own as isolated one-off moments of time. Yet, in our contemporary social media culture, it is permitted, and arguably expected, to assign a greater sense of purpose and meaning to pop icons and moments. Doing so makes us LOL, question and think, feel outraged, and occasionally cry. This process helps us to self-identify, relate, and experience the nature of being human through the emotional connection of pop culture assimilation.

Whether our intent is fun and witty, cute and loving, inspirational, satirical, political, educational, or something else, we contort the original context of pop culture moments so that we, too, can be seen and heard. Essentially, we seek out identity, belonging, and purpose. I'm guilty of this. As a writer, I'm always working at the craft of storytelling, trying to entertain the reader while hopefully educating and enlightening them along the way. In my experience, this is not easy to do.

This said, I'm going to go against my better judgment, sit comfortably with my cringe, and reference a popular streaming Hulu show, *The Bear,* to illuminate a perspective on how we can attain greater prosperity through planet (and people) pragmatism. Let me explain.

When I first discovered FX's *The Bear* on Hulu, I conducted the incredibly unscientific "ten-minute" rule, something I've adopted for my family to quickly evaluate any new show or movie that is streaming. In my unprofessional assessment, ten minutes is just about enough time to get past the first scene, experience some character and plot development, and make the determination on whether you'll invest another 30 minutes or 3 days into the show, particularly if watching with friends or family.

I'll let you know that if there is a show you selfishly want to watch and a family member or two is not yet convinced, the ten-minute rule seems to pull others on board. I think this is because after ten minutes, people feel emotionally invested. Or perhaps the psychology of having a choice in the matter yields buy-in. Plus, if you and others ultimately determine not to watch the show, at least you've ruled out whether the show will be a family affair or an individual pursuit. After all, life is short, and there are a lot of great shows and other things one can be doing with their precious time. So, a few minutes to make a family "go, no-go" decision on a show is a reasonable strategy and a small investment.

If you haven't seen an episode of *The Bear*, forgive me for this pop culture reference. But it cuts deep for me, and I would wager, for anyone who has worked in a restaurant. I had no idea what *The Bear* was about prior to clicking play. I had seen some previews and heard some buzz but had no understanding of the characters or story. All I knew was that the show had loosely cobbled together around a Chicago-based beef sandwich shop. That seemed interesting enough for me to start with. A few minutes into my ten-minute rule watching Season 1, Episode 1, and I was hooked.

The Bear brilliantly touches on something beautiful as it dives into the intensity of the characters' rising emotions, their race against (restaurant service and personal) time, and the desire to find personal joy and meaning in life while simultaneously fighting to survive (in business and in life). I instantly identified and resonated with *The Bear* because I spent a solid ten years of my early life working in restaurants, including working "on the line."

We Are All Working and Living, "On the Line"

For those that may not know, the "line" is a reference to the main cooking area in a commercial kitchen. The "on the line" metaphor reaches far beyond the restaurant kitchen. It's also a fantastic metaphor for the state of affairs of our society, and the intention one gives to the development of their occupation and lifestyle.

There are line workers in the power sector who put their lives at risk every day to ensure we have safe and reliable electricity. There are union workers who picket "the line" in attempts to secure fairer wages and better working conditions. There are the offensive and defensive lines in football, each as critical as the other in determining whether the team will secure a win and successful outcome to the game. There are the courageous entrepreneurs, small business owners, and single working mothers, each respectively "put it on the line" each day, striving to achieve success and a better quality of life and future for themselves and their families. Then there is the long line that we have all experienced, as citizens and consumers, patiently awaiting our turn to renew a license, receive medical care, make that special purchase, celebrate a special occasion, or pay respects to someone who has passed. *Essentially, we are all "on the line," facing most certain personal and professional risks, as we fumble our way through this beautiful thing we call life.*

Back into the kitchen. The art of cuisine, the fast-paced chaotic environment, the nuanced culture and language, the unspoken systems and processes, the creatures and characters, the flexing and compressing of time (i.e., *"every second counts," The Bear*), the pressure points and the elation felt after a good night — work within a restaurant is like a microcosm of greater society. When at its best, the restaurant is like a well-tuned orchestra, everyone has a role to serve, and each person is a link and note that connects the music and amplifies the restaurant's performance, and every customer's experience. When at its worst, the restaurant fails to create an experience, leaving most dissatisfied, disillusioned, and disgruntled. Anyone who has spent time working in a bustling restaurant, whether as a chef, line cook, prep cook, waitstaff, dishwasher, bartender, owner, maintenance staff, host, or in-house entertainer, knows what I'm speaking of.

The experience is more than the food and drink, it's more than the venue and entertainment, it's more than the owner's story and staff. The experience encapsulates the profound sense of connection and community that comes together in celebration, in defeat, and during times of pain and suffering. The restaurant is a place of acceptance and camaraderie. It's a mirror to ourselves and to society, the good and bad, the beautiful and horrid. That's the magic of *The Bear.* The show is about how people define place, how place defines people, and how we are all working to attain belonging, identity, and greater prosperity. When we do it together, as a team, that can be enriching and rewarding. When we try to do it alone, we struggle, if not fail, just as the main character Carmen ("Carmy") tries and experiences as the eccentric chef on *The Bear.*

First Jobs are a Bear: Discovering My Place "On the Line"

One month before my fifteenth birthday, I got hired in my first job as a dishwasher at a local Italian restaurant in Auburn, New York. My older sister was a bus girl and made me aware of the opportunity. After talking it over with my parents, I decided to explore the job. I had no idea what I was getting myself into, but the idea of making a few bucks drove my intrigue and enthusiasm.

My interview was straightforward. I pretty much filled out some paperwork, met the restaurant owners for a conversation, and answered some questions gauging my level of interest, commitment, and character. I was hired as a dishwasher, making minimum wage, which at that time was well below five dollars an hour. But wages aside, I was excited, even as a budding fifteen-year-old, to have a job. However, after my first evening washing dishes, both my enthusiasm and intrigue would wane. Restaurant work is hard. The environment is loud, sticky, and hot. The hours are long. The pay is minimal. During service, the activity can feel maddening.

As a dishwasher, you get to observe a lot. At some point, dishwashers touch every single utensil, pan, rack, cup, bowl, plate, cutting board, and so on. Dishwashers deal with everyone, except the customer. The dishwashing area can be a transitional and transactional zone of warfare and for peace. Often, dishwashers receive the brunt of frustration from all other service providers in the kitchen and throughout the restaurant. I've seen tempers flare and hugs of compassion at the dishwasher battleground as chefs took out their frustration and as waitresses and bartenders apologized for curt comments made during service.

The dishwasher battle zone can be outright nasty. On the one hand, the job is filthy; on the other, you have to be willing to put up with a lot of abuse, most of which has nothing to do with you or your job. The dishwasher is the dumping ground for trash, whether it's tangible food waste or verbal diarrhea spewing from someone needing to vent. In my experience, dishwashers develop, whether they ever wanted to or not, much thicker skin. There are a lot of dirty and tough jobs out there, but dishwashing certainly stands up to many of them.

There is nothing glamorous about scraping food from plates, pots, and pans. I literally went home every evening covered in sweat, food debris, and grease. I smelled terrible. My mom made me leave the sneakers I wore outside. I'd shower and yet could still smell the restaurant on me.

Looking back, dishwashing was a great training ground for a first job. The job was rewarding, and it built work ethic and character. The first paycheck reinforced what all the pain was for.

And, over time, the job became equally fulfilling. As I learned to do the job better — quicker, cleaner, more efficiently, etc.— I became more aware of satisfying *my* customers: wait staff, prep and line cooks, chefs, and owners. It did not take long to understand that, as disgusting and menial as dishwashing was, it served an essential purpose in the overall composition of the restaurant orchestra. Dishes and pans needed to be cleaned, trash needed to be discarded, boxes needed to be broken down and recycled, ovens needed to be degreased and cleaned, walk-in coolers and freezers needed to be sanitized, floors needed to be swept and mopped. All these tasks and many more were standard sheet music that needed to be played, so that the orchestra could hit the crescendo with gusto, night after night.

Ironically, and amid the frenzied and frustrating environment, I began to enjoy working in a restaurant, and soon enough, I began to take on new responsibilities, including supporting prep cooks. Eventually, I would leave that first job and restaurant, only to take another job as a prep cook at another Italian restaurant. Then, a year or two after that, I became a prep cook and then a line cook. Soon enough, I was serving lunch, dinner, and even "late night" for the evening bar dwellers. I learned by watching, listening, and asking questions from those who knew the answer. I also learned by jumping in and doing, modeling the technique of trained chefs and cooks who were educated from the Culinary Institute of America and other fine institutions.

Working in a restaurant, I've burned my feet, hands, arms, and other body parts too many times to count. I've gone to the emergency room for stitches after slicing my hand open with a French knife. I've slipped and fallen on the wet kitchen floor and pulled a muscle or two trying to dump hundred-pound trash containers into the dumpster (usually spilling a good portion down my shirt). Restaurant work is physical and exhausting. But slowly, I gained my sea legs and developed new skills, from safe food preparation to organization and efficiency, cleanliness, cooking, and food presentation. As I saw my skill set grow, so did my pay, also elevating my overall confidence and attitude toward work.

I enjoyed working "on the line" as a line cook. I began working on the frier and pasta station, cooking veal and chicken parmesan and steaming an array of pastas. Over time, I moved "down the line" and onto learning all aspects of grilling a perfect medium-rare filet, baking perfect haddock, and sautéing shrimp piccata, chicken marsala, and bespoke dinner specials.

Serving "on the line" alongside trained cooks and chefs was a privilege and a challenge. A Friday or Saturday dinner rush at the restaurant I worked for could see more than 300 to 350 reservations. That's a lot of table turns and a lot of pasta. Evenings such as those could witness a brilliantly orchestrated performance, or they could have been disastrous. The head chef was usually the conductor, and in my experience, the well-trained and educated chef usually led a tighter performance.

The well-trained, experienced head chef typically sets the tone, pace, attitude, and expectation for each evening. I always found comfort in that, even if it felt a bit dictatorial or militaristic. The ground rules, boundary conditions, and expectations that were provided to me and to others served as a compass and provided a sense of certainty during the turbulent moments of a dinner rush. During moments of what could be construed as chaos, there was focused attention, detail to the task at hand, and an understanding of the bigger picture and objective.

Beyond having a job, I grew fond of the people and place that brought life to the restaurant. Having previously worked as a dishwasher, I learned to respect everyone and the role they served. The work was the work, so to speak, but the diversity of colorful personalities, backgrounds, experiences, and daily mischief and mishaps made for a truly unique experience. My appreciation and love of cooking and the restaurant business expanded, and, for a moment, I considered applying to the Culinary Institute of America to become a chef. I remember thinking methodically about the opportunity and decision. I weighed the pros and cons, but ultimately, I chose another education and career path. All through my high school, undergraduate, and graduate school years, I would return to rejoin the cast of characters and musicians that made up the restaurant circus, symphony, and tragic comedy for a new season of entertainment and work.

My career has progressed quite a bit since my dishwashing days. I've been fortunate to work for public benefits, applied research, academic, manufacturing, engineering, advanced technology, and consulting organizations across a diversity of technical, management, and leadership roles. My career has been an incredible journey, but I still look back fondly on my first job, including the opportunity and learning it provided.

Financially, my restaurant years paid for my first car, a 1985 Chevy Monte Carlo. The experience also paid for my first two years of college tuition and expenses, not to mention so many late teen and early 20s fun entertainment expenses, from golfing to travel, movies, dinners, and so much more. It was the consummate first job experience, teaching me the value of discipline, hard work, finance, organization, management, creative expression, communication, listening, and leadership. The experience challenged me. The people and the environment taught me a lot about myself, including lessons that I needed to be taught, and some things I learned about myself that I did not care for very much. But like most people, I learned more about myself, and I grew. Restaurants can be an unforgiving place to work. It's a job, and you're part of a crew. You either sink or swim. *At the young age of fourteen, I chose to swim.*

Our Society is Far From Attaining Our First Michelin Star. Can a Purposeful Pursuit of Planet Prosperity Get Us There?

Okay, if my "on the line" restaurant experience and *The Bear* metaphors have not turned you away yet, I'm shocked, but grateful. Thanks for sticking with me and letting me try to bring all this home. Fast forward, and my "ten-minute" show rule for *The Bear* resulted in me consuming hours of watching the balance of Seasons 1, 2, and 3. I had no idea of the deeper trauma, drama, and journey that the show would encapsulate. It was all well worth it, at least in my assessment of a pretty decent show.

As one of the threads of the show, the main character and gifted chef, Carmen ("Carmy"), chose to pursue a Michelin Star, the epitome of personal achievement and restaurant excellence. Having previously worked at award-winning Michelin Star restaurants, Carmy knew, at least from one perspective, what that image of greatness looked like. During Seasons 1 through 3, the show captures the transformation of Carmy's dead brother's beef sandwich shop into a flourishing hip new restaurant. To get the restaurant into Michelin shape, Carmy's vision required a completely artisan and bespoke menu, flawless execution, impeccable service, deep commitment to structure and systems, and intolerance for anything but excellence.

Carmy's obsessive pursuit of a Michelin Star, while an image of greatness, also cost him a great deal in his personal relationships and the joy that originally fueled his creativity and enabled his gifted abilities as a chef. Carmy was at war with himself, fighting his own demons and perceptions of greatness, much of which had been self-imposed and constructed from his past relationships with family, friends, and those who trained him professionally. Ultimately, Carmy had to make a choice and determine how to relinquish the source energy that fed his joy, while also navigating the true purpose of leading people, and toward a new definition of prosperity and success. For those who haven't watched *The Bear,* I won't give away any surprises here.

The point is that our journey as a society and as individuals are both unique and interlocked. Working to achieve personal prosperity at the expense of the greater good may serve a few, but this mantra of selfish success has caused and is continuing to create enormous damage to the environment and to the greater well-being of society.

The year 2024 marks a pivotal time in history. We have access to incredible research, technology, infrastructure, and resources like no other time in human evolution. We arguably have the best of the best tools, facilities, ingredients, and team to do exceptional things that better the fate and future of humanity. We can cure disease, launch rockets, and have artificial intelligence do our homework. Yet, we remain mired in doing things the same old way, and in some cases, reverting social progress backwards by decades and more.

We're like Carmy, gifted and skilled, well-trained, but also alone and broken. We're tethered to such an ideal of perfectionism for greater prosperity that we have forgotten that it is, and has always been, about the pursuit. We must not take anything as a given or for granted. We may or may not achieve a Michelin Star in our pursuit of greater prosperity. But we must not forget that the prize, the win, the accomplishment will only happen when the entire kitchen staff (society) serves and performs with excellence and in a united way.

Points on Pragmatism

- *The pursuit of prosperity is challenging during this time of swift technological change, social unease, economic uncertainty and turbulence, and environmental challenges. The planet and our prosperity are literally, and figuratively, "on the line," as we determine our fate and future as global citizens and consumers. To achieve personal and societal prosperity, we must recognize, respect, reinforce, and reward the fact that we all have a role to serve.* ***Prosperity only occurs when everyone contributes and is treated equitably and with dignity.***
- *Serving society, or living, "on the line," can be exhilarating, particularly when risk can be proactively identified, managed, and mitigated.*

Although the future can feel uncertain, we can positively effectuate change and choose to minimize fear and exemplify excellence in how we unite with compassion, courage, and resolve. We must recognize that our mindset is a choice, as are our actions, as we pursue prosperity. A clear mind can yield focused results, particularly when it is reinforced with a positive attitude. ***If we are grateful, eager, and willing to put in the hard work, we all will learn, grow, and prosper together.***

- *Life is a bear, but it can be tamed. Society is yearning for greater connection, beyond the feigned and fickle likes and shallow emptiness of our pop and social media culture. We must take it upon ourselves to tear down fictional barriers in how we choose to communicate, collaborate, learn, live, and grow.* ***Taming the bear requires all of us to look in the mirror to question our own behaviors in the face of our true beliefs and values.***
- *Planet pragmatism is about taking practical steps toward attaining a higher quality of life and enriching life's journey now, in the present moment. It's about tapping into the innate wisdom we all have, a desire to be seen, to have identity, to feel safe, to have a sense of belonging, and to discover and serve others with intention and purpose.* ***Planet pragmatism is not about chasing shiny objects or achieving Michelin Stars. It's about celebrating the vast richness that life has to offer, and which is already abundantly available to us.***

13

PURSUING A NEW PROSPERITY THROUGH PLANET PRAGMATISM

Once a beacon of prestige and freedom, the classic metaphor for pursuing and attaining the American dream has eroded. Decades in the making, the demise of the American dream represents a confluence of economic, environmental, and social forces that have been eating away at people's quality of life and their pursuit of prosperity. Although the romanticized American dream has faded, a new generation of Americans and global citizens is seeking to define their fate and freedom toward a new prosperity. Fed up with big business, big brother, and big tech, citizens and consumers feel as if they are trapped living within a world of deception, delusion, and distrust — some might even associate the dissolution of the American dream to the advance of the dystopian state captured within George Orwell's *1984*.

The American society, and democracy at large, has long had to fight for survival. We've been shaken by a steady state of social change and civil unrest that some historians and pundits postulate as the growing pains America and Americans have had to go through — as democracy and freedom are never simply afforded, they must be intentionally pursued and protected. The origins of a new prosperity are underway. Guided by redefining prosperity (the underlying "why") and through principles of pragmatism (the "what and how" of the roadmap), people have shed notions of enslaving themselves to institutions and systems that permeate false beliefs and narratives around freedom and prosperity. People are reclaiming their virtue and assuming control of their future. While a rebellion of ideology may ensue, if not a more pronounced rebellion that could arise, my sense is that individuals and our broader society are yearning for a greater sense of certainty, belonging, identity, hope, and love. This is not to say that we will not continue to see social uprisings or heated debates in the coming months and years. Given the climate we're living in, we are literally and figuratively, as Thomas Friedman's book title connotes, living in a *Hot, Flat, and Crowded* world.

Truly, the divisive and heated political temperament seems to rise like the planet's temperature. Our resources are limited, yet we continue to be wasteful. Although the corporate sustainability movement is visible and billions of dollars flow toward "sustainable growth," billions of people around the world live in dire conditions, fighting for their survival. The movement toward a new prosperity has mobilized. In the next decade, this movement will rethink and redefine our notion of wealth, ownership, status, and power. In doing so, this movement will become a new hope toward a more peaceful, just, equitable, dignified, and sustainable society. This chapter outlines a few of the principles of pragmatism that can be explored toward advancing a new prosperity.

Putting the Principles of Planet Pragmatism into Practice

Embracing and practicing the principles of planet pragmatism in pursuit of greater prosperity equates to responsible risk management combined with a proactive posture for solving real-world problems through innovation.

Over my 25-year career, I have witnessed many recurring pendulum shifts in politics, policy, technology, leadership and management, societal values, and institutional priorities. Such shifts are indicative of and signal change. Often, these shifts simply ramp up and roll out of previously attempted innovation. Solving real-world problems with innovation requires an appetite for taking calculated and mitigated risks. The risks can be defined against any number of parameters: economic/financial, political, institutional, societal, operational, infrastructure, environmental/climate, public safety, national security, and so on.

Weighing the benefits of "solving for X" (i.e., decarbonizing the economy, climate adaptation, addressing water scarcity and food insecurity, infrastructure security and resilience, sustainable production and consumption, etc.) against the known, known-unknown, and unknown risks is challenging and, from this author's perspective, remains elusive. Herein lies the benefit of planet pragmatism. We don't have all the answers, data points, or intelligence to derisk every decision that we make. However, if guided by our collective wisdom, shared knowledge, and societal values, we can certainly construct a risk-based framework that can optimize outputs for environmental, social, and economic good while minimizing known risk and unintended impacts. Through a preventive, predictive, and proactive posture on how we bring innovation forward (in all aspects of new ideas through solutions, not just technological), we can optimize the resources available to us today, as we make informed decisions to "solve for X" now and into the future.

If you think of this as a process or system, it is not linear, nor is it self-contained. Planet pragmatism requires us to constantly be listening, learning, and engaging to challenge the old and invent the new. Planet pragmatism requires ongoing communication, continuous iteration and improvement, and optimization of all resources (i.e., natural resources as well as financial, technological, human and talent, data and digital).

If principles for planet pragmatism are to yield the progress toward prosperity that we seek, we must always be actively sensing our world, asking tough questions, seeking answers and truth, and willing to act on our values and moral convictions. Having principles on paper is not enough. Principles must be acted upon, tested, and evaluated. To be effective, principles must also be embraced and lived as a matter of practice and faith. Practicing planet pragmatism prepares us to challenge, adapt, and evolve our principles in step with the wisdom and knowledge that are revealed over time. We must be willing to listen, learn, and adapt. Change is not something to fight with might; rather, it is a force that can, when guided by principles of planet pragmatism, yield purposeful innovation and greater prosperity.

Take, for example, the evolution of our modern energy sector. Around 3,000 B.C., the Babylonians (modern day Iraqis) were one of the earliest civilizations that we know exploited oil as a resource. The Babylonians[106] used crude oil, which bubbled to the surface of the Earth, to waterproof their boats. Archaeologists also discovered that the Babylonians used oil as mortar in building construction. Oil has also been found as one of the substances used by the Egyptians in mummification. In 600 B.C., the Chinese discovered oil and transported it using bamboo piping as infrastructure.

The industrialization of what we know of as the oil industry did not happen until well after the "black gold" was discovered in Titusville, Pennsylvania, in 1859 by Colonel Drake. Fast forward 166 years, and our modern economy operates exclusively on oil, gas, and their associated by-products. Industrialized production and use of the black sticky substance transformed it into the most ubiquitous global energy source. Oil provided, and continues to provide, the means to fuel and power a burgeoning U.S. and global economy. Oil had a great run for over a hundred years.

Then in 1973, the "oil embargo[107]," initiated by the Organization of Arab Petroleum Exporting Countries (OAPEC), created a precedent, and new fear, when oil and gas were politically and economically weaponized as a global trade tool. Shortly thereafter, the U.S. government and states, including California, began enacting the nation's first energy efficiency standards for transportation and transportation fuels. A couple of decades earlier, the U.S. had already begun to explore residential housing efficiency standards.

The 1973 oil crisis was a wake-up call for the U.S. and all global industrial nations whose economies had become reliant on foreign oil and gas. Resultingly, the 1973 global oil crisis resulted in a pendulum shift. Those who controlled energy, and specifically oil, now controlled political and economic power. This realization resulted in a sea change in the perception of oil as a primary fuel source. In just over one hundred years after Colonel Drake's discovery marking the global industrialization of oil, society began a transition toward a new era of energy production and use. This new era was marked by the pragmatic necessity for energy efficiency, energy and national security, and environmental protection. The once myopic focus on exclusively optimizing oil production and use began to shift toward energy conservation, efficiency, and production and use of alternative energy sources.

Fast forward a few decades. Oil remains the predominant source of energy in the U.S. and throughout the world. In fact, global demand for oil has increased since the 1973 oil crisis. Plagued by concerns over climate mitigation and adaptation, today the oil and gas sector and the broader energy sector continue to explore and pursue alternative energy sources and pathways for decarbonized fuels. Since the 1970s, the energy sector has developed numerous advancements. Breakthroughs in physics, materials and chemicals research, digital solutions, renewable energy, and engineering have led to advancements for primary energy sources (i.e., alternative fuels, nuclear, wind, solar, geothermal), and for energy system dynamics such as energy storage, transmission, and distribution. (i.e., battery storage, lower carbon fuels).

However, no singular technology has been deployed anywhere near the global scale of oil.

For the past fifty years, the energy pendulum has been swinging back and forth, bringing forth new ideas, waves of innovation, and slow but steady progress toward decoupling the world from the stronghold of its oil-centric economy, giving rise to one that is more energy diverse. Right now, we are experiencing another major energy pendulum swing. Although in contrast to President-elect Trump's inaugural speech[108] touting his policy directive for energy security as "drill, baby, drill!," in 2025 the ongoing critical role and future for nuclear, hydrogen, geothermal, battery energy storage, energy efficiency and natural gas have been elevated, yet again, as critical technologies, cleaner fuels, and energy solutions that can deliver an affordable, clean, and reliable energy future. Interestingly, these technologies have been on our radar and undergoing research, technology demonstration, market deployment, and scaled integration with infrastructure for decades.

Energy pendulum swings have occurred several times throughout my short career. In the 2001 timeframe, I took a brief ride in one of General Motors' hydrogen fuel cell prototypes when I was an energy analyst working at the New York State Energy Research and Development Authority (NYSERDA). In 2008, I worked closely with an engineering and applied research team on a NYSERDA-sponsored hydrogen storage, fueling, and internal combustion engine (H2-ICE) deployment at Rochester Institute of Technology (RIT). RIT's campus safety operated three H2-ICE vehicles as part of its fleet. I worked with a team of engineers, applied research scientists, project and program managers who sought to better understand the efficacy of hydrogen fueling stations, vehicle performance, safety requirements, economic costs, and other key variables. These examples represent real-world deployments of advanced energy technologies that have been undergoing research, testing, and deployment, and further development for years. Now in 2025, it feels like we're going back to the future, and what was once old, forgotten, and perhaps written-off technology is now again en vogue.

As the U.S. and the world butt up against a new era of rising energy demand, social need, and climate related risks, alternative energy pathways are once again being explored. From a pragmatist perspective, our global energy economy should never have been about one fuel, yet we made it so. Ever since, we have been working against the strong tide of oil to diversify and decarbonize global energy sources. Not all energy sources are created equal, nor are they the perfect match for all energy uses. The intermittency of renewable power from wind and solar is often criticized, for example. The molecules that comprise oil happen to be energy-dense, and as a liquid, oil has proven to be relatively easy to design infrastructure around. Accordingly, as an energy source, oil has been the low-hanging fruit and path of least resistance for humanity to design an entire economy around. We can and must do better.

Although it has historically been cheap, easy, and convenient to scale up the production and use of oil as a primary energy source as compared to alternatives, the inconvenient reality for the planet and for society is that the social and environmental externalities of this fossil fuel now outweigh its nominal economic benefits. Today, we have available to us the necessary wisdom, intellect, and know-how to redesign the global economy with energy sources and infrastructure that align with our values and needs. By pursuing planet pragmatism, society can optimize resources, reduce waste and pollution, and deliver a better quality of life for people and within the limits of planetary systems. Oil has been our go-to fuel for hundreds of years. We are just now working to transition our energy and economy in tandem, leveraging principles of planet pragmatism in a way that delivers us greater prosperity, while protecting human health and the environment. An energy transition is underway. The decades between 2020 and 2050 will witness fundamental changes in how we produce and use energy and power our modern society. Traditional and essential operating requirements for safety, security, reliability, quality, and affordability will shape the next three decades of energy transition. So, too, will new market economy requirements, as defined by principles of planet pragmatism, such as the diversification, decentralization, decarbonization, digitization, democratization, and derisking of energy.

President Trump won the 2024 U.S. election. As the calendar flips to 2025, the earth beneath our feet is quite literally moving, and significant change is afoot. Trump's transition back into the White

House for a second term can only be characterized as swift, intense, and dizzying. Within days of taking the oath of office, the President signed more than fifty Executive Orders[109] (EOs) that swept across a diversity of social, economic, and environmental concerns including AI; energy security; diversity, equity, and inclusion (DEI); transgender rights; trade and tariffs; birthright citizenship; U.S. withdrawal from the Paris climate agreement; and much more. What's more, President Trump and his in-bound administration acolytes continued their campaign of shock and awe with a litany of newsworthy actions including the deep budget cuts proposed by Elon Musk and the newly formed Department of Government Efficiency (DOGE), the renaming of U.S. landmarks and territories such as the Gulf of Mexico to the Gulf of America, and the barrage of international posturing including Trump's call for Canada and even Greenland to be recognized as U.S. territory. President Trump's second-term scorecard for the first one hundred days in office will be one for the books. Only time will tell what repercussions and opportunities emerge from President Trump's second term. For now, the shock and awe are leaving most U.S. citizens, and arguably many global citizens, anxious, confused, and concerned. Prosperity is a long-term objective, but its pursuit, much like democracy, must be fought for daily. As the pendulum of the evolving world order swings, the fate and freedom of our global society hang in the balance.

Fundamentally, our collective future hinges upon a verdant and vibrant global environment that delivers ecosystem services for humanity, including clean air, fresh water, healthy food, and climate resilience. The power-seeking politicians who seek to rule by breaking up the old rules of governing are simply exploiting civil society during one of its weakest points. We may not yet fully realize it, but these are highly vulnerable times. Climate risk is escalating rapidly. Our food systems, housing, transportation and infrastructure, healthcare, and much more are all being impacted by an increasingly fragile natural environment. Mass migration of people is not something that happens "over there," in developing countries. Climate risk has escalated. Here in America, the mass migration of people due to wildfires, water and food scarcity, and the devastating impacts associated with hurricane storm surge, strong winds, and even winter storms and ice. Our environment has changed. Storm events are more severe and less predictable. Society is in a climate crisis.

Those in power right now know and understand this, even those who pull out of climate accords and who do not have climate as a top-tier agenda item. Such politicians are not really leaders of the people concerned about the common good. Rather, they see humanity as what it is in this moment, vulnerable and unprepared. Thus, they have weaponized technology, disinformation, and a planet in duress against the people, pitting the continued financial success of a few over the common good for all.

To be clear, a war is being waged, and most people do not even see or understand that they are part of it. More than half of the world's population is living each day to survive. Then the next forty-five percent of the population is working and earning a living to survive under the guise that they are thriving. Then, there are the select few, less than five percent, who are surviving well and trying further to thrive during vulnerable times. Finally, less than one percent are those who are waging economic and environmental war on the rest of society. They are the rule makers and rule changers. They don't see the world as you and I do. They are not our saviors or peacemakers. As citizens and consumers working together with common sense for the common good, We can only be successful in winning this war and achieving the freedom and prosperity that we so desire by acting together with common sense for the common good.

Surely, the institutions, systems, and inefficient market forces that delivered a global economy over the past century have served their purpose. Undoubtedly, nation states and civil society will benefit from innovation not only in the palm of our hands, but also in the governing institutions that are supposed to guide us with principled leadership. We mustn't forget that there is no substitute for strong governance, principled leadership, and planet pragmatism. If the environment crumbles, society crumbles. If society crumbles, the precious one percent that wage war will succumb to their own fate.

The balance of this chapter lays out, in simple language and viewpoints, the building blocks of ideas that can feed the development of principles for planet pragmatism. My hope is that the thesis of this book is explored earnestly by students, researchers, futurists, consultants, sustainability practitioners, engineers, teachers, executives, and all others working to reconcile their value and place in a rapidly changing world. No one person can singularly shape principles for planet pragmatism alone. This book and its ideas are meant to provoke deeper thought and introspection, along with healthy debate and ideation.

Right now, a war for our minds and our existing and future prosperity is raging. We inherently know that systems are broken, the planet is suffering, and our ideals for peace, freedom, and prosperity are not being achieved. We must band together and create our own playbook for planet prosperity. In doing so, we will reclaim our individual and collective rights, freedoms, and power. Giving in or up on our dreams for prosperity is not an option. To do so would only further invigorate those who seek control over our future. We have the necessary ingredients for creating a safer, more sustainable, and benevolent society. Putting the principles of planet pragmatism into action is simple. All we need to do is join forces, define our common goals and principles, create our planet prosperity playbook (collective action plan), roll up our sleeves, and get to work.

Getting the Foundation Right: The Building Blocks for Planet Pragmatism

Wisdom is useless unless we put it to use. Essentially, we need to learn from the past. And we need to stop doing stupid things on purpose. We will only achieve a higher quality of life and greater prosperity when we allow the full wisdom of society to guide our collective action and resolve. There are many things we do and get wrong, and yet, we do them all too well. For example, we need to stop environmental degradation in all its forms, and at the hands of human behavior, and our overconsumption of natural resources. Waste of all kinds can be mitigated by rethinking the purpose behind our products and services. Bringing a pragmatic, system-oriented, and holistic mindset to any problem can yield better outcomes. This requires us to go beyond *what's in it for me?* To manifest greater prosperity and create a more safe, resilient, and sustainable world, we need to think about our individual roles and responsibilities, and how we impact the world around us.

Let's show some humility and restraint, especially when it comes to advanced technology. Just because we can, doesn't mean we should. This adage seems to keep playing out for humanity, time and again. We cracked the atom. Then we found the God particle by carefully observing the collision of atoms. We sequenced DNA and then found a way to edit DNA at our will (talk about a God complex!). And now we have unleashed the power and potential of AI in society. Across humanity's technological prowess and musings, the proverbial toothpaste has spewed out of the tube, and at this point, there is no chance of us putting the toothpaste back in. We must live with the beauty and bounty and the undetermined and unintended consequences of our ingenuity. Human tinkering has led to some remarkable inventions, innovations, and technological progressions. It has also led to many social, environmental, security, and economic disasters and ongoing challenges. There are technologists and economists who believe that externalities are the utilitarian social and economic cost associated with advancing society. We may not be able to preempt every negative externality that could arise from the advance of technology in society, but we can and should leverage our knowledge, ethics, and morality more prominently in how we introduce innovation to the world. Look no further than the use of generative AI and social media. Our society has unleashed technological capabilities that can, for those that choose to wield them irresponsibly, distort fact from fiction, spark division, and erode the fabric of trust within society.

Nobody wins in war. We must end unnecessary conflict and acts of aggression that cause undue harm to people and planet. From a pragmatist perspective, conflict is natural and necessary. With more than 8 billion people living on the planet, increasingly competing for finite resources, attention, and significance — conflicts are bound to happen. However, not every errant disagreement, differentiating point of view, or abhorrent gesture among us needs to result in a conflict of aggression. Whether it is within our home, our local community, domestic border, or on the world stage, we need to have constructive conflicts that value people and planet, and which seek peaceful and productive resolution. We have a long way to go. The continued advance of humans will require that we grow, not only through our advances of technology, but in leaps and bounds, emotionally, spiritually, and in how we enact *common sense for the common good* through planet pragmatism.

Disagreements can drive creativity. Although conflict and acts of aggression are inherently destructive and unsustainable, we must seek to differentiate between friendly and unfriendly conflict. The planet and people across multiple generations have been shaken by continuous conflict defined by differences in geopolitical, cultural, religious, socio-economic, and other ideologies. Understandably, most people shy away from conflict and try to avoid disagreements. To avoid conflict, too often people seek the path of least resistance — in relationships, careers, family dynamics, community development, politics, and life in general. While it is true that disagreements can lead to conflict, we should not cower as a default for dealing with our differences. Our diversity is an asset to be cherished, celebrated, and consecrated. Our differences shall not divide us, rather, they represent principled perspectives that should be valued and the creative means by which we can explore, evaluate, and solve complex problems together. In a world that Thomas Friedman aptly characterized as *Hot, Flat, and Crowded,* the 8 billion people (and counting) living today are bound to have a different perspective on just about everything.

Really, it's a wonder that our global economy, social institutions, local communities, and family households operate on any level. Think about it. Eight billion spirits, minds, egos, hungry bellies, and visceral needs are a lot of daily decisions, demands, and disagreements. When we flip the script that many people have lived or been taught, we realize that disagreements can be a source of creative energy and ideas that can yield better results than any one overpowering point of view. Thus, it is imperative for our generation to guide each other and the next generation with the skills, tools, education, and mindsets that embrace conflict resolution, trust-building, tolerance, diversity and inclusion, constructive criticism, and critical thinking. These traits and skills build character, hone sound judgment and critical thinking skills, and result in a more well-rounded, systems-based, and holistic mind that elevates intelligence to use common sense for the common good. Sure, we all have innate needs, egos, and selfish pursuits — but if we are to be successful in pursuing and achieving prosperity in a rapidly changing world, then we need to access and operate with a higher sense of self, as well as collectively and creatively, aligned with principles of planet pragmatism. When we finally place the appropriate value on our diversity, we will then recognize and validate that our differences create tension that forces us to ask deeper questions, seek honest truths, and discover how to catalyze creative and more collective solutions. Creative processes reinforced by an intentional call-to-action agenda allow participants to reevaluate complex problems from multiple dimensions. In doing so, proposed solutions tend to be more holistic, thoughtful, and less likely to perpetuate inequalities and inefficiencies from unconscious bias or systemic institutional barriers.

Scarcity and necessity beget innovation, but remember, complex problems don't necessarily require complex solutions. Our future prosperity will be predicated on our ability to adapt to change and ground ourselves in shared values and principles for planet pragmatism. Our planet is incredibly abundant and diverse, teeming with life. Yet the geopolitical power dynamic and global economy have most people living with real scarcity of resources, or a felt sense of never having enough. The planet has enough resources to provide all people and cultures with a sustainable existence and high quality of life. Our planet's resources are not optimized to this outcome, however. Rather, nation states view resources as competitive options and commodities that can be converted into wealth-creating assets. In doing so, we've created competition over the planet's commons, and subsequently have manifested models for unsustainable production, consumption, and conflict. This geopolitical paradigm will not change quickly. Thus, our ability to remain adaptive, inventive, and pragmatic in our pursuit of greater prosperity is essential. Unfortunately, competition for natural resources is becoming fiercer. The geopolitical landscape is shifting, and nation-states have become more aggressive and authoritarian in their disposition toward global trade, national security, and future growth. Instead of seeking common sense and collaborative means to pursue the common good, many nations are pitting themselves against one another in a global race to control rare earth minerals, oil and gas, and other energy resources, global trade routes, and regional cultures and people. The competitive posturing of nations is highly likely to intensify in coming months and years, brought on by social, economic, and environmental pressures tied to climate and ecosystem resilience risk. Our global economy is complex, perhaps too complex. Too often, the solution for "Solving for X" is an overly engineered, technological solution. Need to remove excess carbon dioxide (CO2) from the atmosphere? Direct air capture (DAC) can be an engineered solution for that. Need to have less costly human labor, less human error, and greater profitability? Artificial intelligence (AI) can provide an engineered solution for that as well. The full economic, social, and environmental costs, trade-offs, and externalities are unfortunately not factored into many of the engineered "solutions" that make their way to design, development, production, and market deployment. In short, we often perpetuate market inefficiencies and failures as opposed to solving them because the underlying structures and mental model of the problem (the root cause) are never addressed. Rather, we tend to overengineer fixes based upon the patterns and trends that are discernible with existing data, and the real-world events that are observed, but often obfuscated by those controlling politics and power. This is why current markets reinforce human-built systems at scale, to try to control carbon emissions as opposed to investing more wisely, efficiently, and effectively in nature-based solutions that can yield additional benefits as well. Because we don't challenge the underlying structure and belief system of how and why humanity pollutes in the first place, we fail to change our behavior and guide our future prosperity in a principled way. Principled prosperity requires us to ask more informed questions and dive deeper into the ways the world operates. For planet pragmatism to be an effective tool in our playbook toward prosperity, we must be willing to challenge the status quo and the belief systems underlying the mental models that have given rise to our institutions, governing bodies, economy, and society. Trying to bring about sustainability, peace, and prosperity in a world that has unsustainable and unpeaceful systems in play and in power is a fool's game. At some point, we must stop playing the game and challenge the preexisting beliefs, norms, principles, and rules that got us to where we are, and question if they can get us to where we desire to go. This process is unnerving, as it will challenge the worldview and mindset for most people. The wheel does not need to be reinvented, so to speak. We must simply calibrate our principles for planet pragmatism with the times that we are now living in, and toward a future that foretells challenging times ahead. Often, the most obvious, simpler, and yet principled solution yields a more sustainable and prosperous outcome.

Prepare for the future but live for now. Life happens in the moment. Yet we often live for or in the past or delay our gratification into the future. Many of the things we occupy our time with have a delayed gratification — exercise, saving for a home down payment or retirement, attaining a new professional certification or academic degree, even preparing and cooking a healthy and delicious meal. Whether it is minutes, hours, days, weeks, months, or even years — the impact of many activities we willingly embark upon in life may not be felt or seen immediately; rather, it is experienced in the future. This is why it is important, for example, to find joy in the process of life.

Protect the planet's ecology and resources as if life depends on it, because it does. Prosperity does not exist in the absence of a verdant and vibrant planet. While the close relationship between planetary and human health should be well understood, we do not behave accordingly. In fact, humans continue to live outside the boundaries of the planet's carrying capacity. We are depleting natural resources and inflicting environmental damage at a rate that exceeds the planet's ability to rejuvenate.

Redefine and rebuild communities that foster quality of life and "real" community. In the past four decades, the idea (and sense) of community has shifted in America. Once idyllic hubs of "main street" economics and livable communities, many American cities, towns, and villages have succumbed to the forces of continued suburbanization and urban sprawl, urban gentrification, the exodus of small business, the rise of big box retailers, and our current culture characterized by free shipping, [un]social media, and remote work. Across America, there are cities, towns, and villages that exude culture and have a strong community. But we cannot dismiss the fact that community is something that needs to be intentionally and continuously nurtured, reinforced, redefined, and celebrated. Just like the world we live in, our communities are ever-changing. How we manifest community serves as a foundation to the external world, yet the external world also actively shapes how we engage in our community. Community is what we make of it. Communities are shaped by many factors, many of which are interrelated and/or self-reinforcing such as local and regional planning, economics, demographics, history, industry, access to education, the existence of faith-based institutions, topography and geography, weather, access to natural resources, environmental pollution, access to healthcare, access to broadband and other communication services, access to agriculture, grocery stores, and food services, cultural heritage history and sites, access to nature and parklands, access to energy and energy services, transportation services, and so much more. When you break down all the factors that support a community, it gets unwieldy quickly. Ultimately, community is about the people, the culture they create, and the infrastructure that people invest in to create the quality of life that supports and sustains the community.

Stop blaming others and start taking responsibility, accountability, and ownership over our pursuit of prosperity. It can be quite convenient in today's world to point the finger at big business, big government, big technology, and any other person, faction, or circumstance that makes us feel that we have an underlying disadvantage in the pursuit and attainment of prosperity. Every person is born into this world with unique circumstances. We each have a unique fingerprint that makes us who we are, including our views, biases, and philosophy on the world around us. Initially in our lives, there is much outside our control: where we are born, who our parents are, the DNA within our body, the early childhood environment and experiences that shape our intellect, the geopolitical climate, and so on. But as we age, we begin to assume greater autonomy over our lives. The autonomy I speak of here stems first from our innate wisdom, then as manifested by our thoughts, and finally carried out in our behaviors and actions. Many people feel trapped, fearful, insecure, lonely, lost, unloved, unhappy, and spiteful because they have not been able to break free, not from the systemic biases of society, but first from their own self-inflicted limitations. We are all handicapped by something. That fact is, there is a part of our human experience that can, with a shared dignity, bind us in spirit and in life. There are people all around us, in our homes, communities, and throughout the world, who suffer from pervasive persecution and a litany of social, economic, institutional, and systemic injustices that continue to plague society. Each of our prosperity fingerprints is different. But wherever we begin does not have to define who we are or where we are going. Where we begin this journey called life is not where we need to end up. The journey will be different, but everyone can change their circumstances. Doing so will be more difficult for some than others, but it is not impossible. Doing so requires that we shift our thinking and behaviors beyond a mentality of blame to a focus on action, accountability, and ownership. What can I do right now, today, to make a change in my circumstances? Even small changes in our thinking and behavior can accumulate and enable a brighter and more prosperous future.

Redefine wealth and prosperity on your terms. To live simply is to live wealthily. True wealth is subjectively measured by one's quality of life and by integrating the dimensions of physical and emotional health, happiness, and spiritual maturity. A wealthy person may or may not have financial wealth. Financial wealth, measured as a societal construct, does not always translate to personal happiness and well-being. A wealthy person can be someone who needs and wants nothing and finds joy and happiness in giving. A wealthy person may be someone who is fully content with who they are and who is at relative peace with the world around them. Wealthy people do not fight against their sense of identity, belonging, or purpose. Their life is enriched, and they enrich the lives of others, by knowing who they are. The pursuit and accumulation of external status through material possessions, financial wealth, friends or relationships, personal accolades, or social mobility do not drive the actions of wealthy people. Wealthy people may have an abundance of these success indicators, at least through the lens of what society views as success. But wealthy people are wealthy even in the absence of any of these external elements.

Raise the floor and the ceiling on sustainability education and upskilling. Believe it or not, right now, sustainability is not a topic that is widely taught in K-12 education, and U.S. teachers and education lag behind their global peers. As a researcher, writer, educator, and mentor, I find this one of the most concerning observations. If we are not proactively providing integrated education to our youth on human understanding of planetary systems, including our role in aiding or deterring sustainability, we are missing out on one of the foundational and most pragmatic steps to improve our quality of life and prosperity. Our youth represents the future of humanity. They will step into this world and its state of economic, social, and environmental challenge and promise. We must give them the tools to succeed and teach them not to make the mistakes of the past. We must provide them with the principles, wisdom, and confidence that they will need as they inherit the responsibility of moving their generation forward, and that they can do so by effectuating positive change. We should not fear or over-intellectualize what we can do right now to provide the necessary level of education, training, and transfer of knowledge and wisdom to our youth. Each day that we fail to act, we lose precious time and potential for planetary pragmatism to take root and to take effect. According to a report[110] by the Smithsonian Science Education Center and Gallup, teachers in Brazil, Canada, France, and India are more than three times as likely to "say they have the necessary support to incorporate topics on sustainable development, like climate action and clean energy, into their curriculum," compared to their U.S. peers. Citing lack of resources, instruction and preparation time, instructional materials, expertise, and professional development — less than 50% of US teachers include sustainable development topics within their curriculum. For example, "just 31 percent of U.S. teachers say they talk about responsible consumption in class — compared to 83 percent of teachers in Brazil." Further, the Smithsonian study noted that about a third of U.S. teachers "say that sustainability topics are not appropriate for the grades they teach — a concern most commonly shared among elementary teachers," and "about two-thirds of teachers say that sustainability does not fit into what they teach.[111]" The U.S. education system, K-12 through higher education, is struggling. Much like we see in other segments of business and society, the transition and transformation of our education systems have not kept pace with the shifting realities of society. This is not to say that every teacher, administrator, or school is not doing what they can to evolve or innovate. There are many amazing teachers, schools, and transformational examples that exist. Pan out from some isolated and well-funded schools, however, and the situation is much bleaker. Schools are underfunded, understaffed, and under-resourced to be effective agents of change. They stick to what is tried and true, a rinse-and-repeat doctrine of education delivery that hinges on a siloed approach to learning. Sustainability offers us an opportunity to breathe new life, fresh perspective, and innovation into our educational system and the desired outcomes for our next generation of leaders.

Go beyond "conscious consumerism," and ask yourself what you really need to be happy, content, fulfilled in your life. Conscious consumerism has a great ring to it. Many sustainability practitioners like to say it, as the alliteration just lets the words roll off the tongue. Conscious consumerism is a great tool and tactic in humanity's fight against unsustainable and mindless consumption. However, as I've mentioned before, we cannot consume our way to a more sustainable lifestyle, future, or planet. I'm all for conscious consumerism, grounded by a "stop, drop, and roll" framework. The next time you're ready to purchase anything, stop! Evaluate your next move as it will have implications for your pocketbook, finances, possibly your relationships, and certainly the impact you are choosing to have on society and the planet. Every time we purchase anything — our lunch, groceries, a Netflix show, additional cloud storage for our amazing vacation photos, the lip-gloss, the toy for the child, the bag of dog food, the new car, cryptocurrency or stocks, a gym membership — all of these and all consumer purchases directly correlate to a social, economic, and environmental impact of some kind. It can feel and certainly is overwhelming to evaluate every single purchase and put us through some type of "green guilt" for simply being human. If they are doing their job correctly, consumer products and services should align with our principles and deliver us some form, if not multiple forms, of value. Conscious consumerism is about being aware of our buying habits and behaviors, and making sure we are not operating in "hypnotic consumer mode," simply buying something to buy it. I've personally been there and, on occasion, I must question myself whether I really need whatever it is I'm about to purchase. So, I stop before making any decision, as a momentary check-in with my needs and true motivation. Then, I "drop" the card, the phone, the cash — or whatever modality of financial transaction I'm about to make. I use this as a second opportunity to review this decision against my financial situation. Am I making a significant financial purchase, for example? If so, do I have the financial resources? How might this purchase impact my monthly financial budget? How will this purchase, if it goes through, impact my credit score or lendability should I want to take out a loan in the future? If this is a sizeable financial transaction, and if I share my finances with a partner, did my partner and I discuss this transaction? If not, will and how might this decision impact my relationship? As you can see, the stop and then drop provides some clarity to you, the consumer, to try and ensure you don't mindlessly give in to our overstimulated consumptive society. I like to say, "It's not your fault," and yes, consumer pressures are real. But you can control your impulses with these simple steps. Finally, if you need to, be prepared to "roll," which means get the hell out of there or the situation. Roll away, on your bike or scooter, in a car, train, truck, or bus. Whatever the fastest beeline for the door is, take it. Walk off the moment. Don't worry, nobody saw you stop, drop, and roll. If they did, they are nodding their head with confident admiration and considering initiating the equivalent of a consumer slow clap moment that crescendos with the entire restaurant, mall, or car dealership shouting in an uproar that you did it! We've all been there, unconsciously swiping our finances away, let alone even thinking about the ramifications that are brought by this lazy and irresponsible form of consumerism. Take back your control and move beyond simple conscious consumerism by practicing the "stop, drop, and roll" method next time you are in purchase mode.

Get back to nature, get back to basics. The quiet calm and tranquility of a morning walk before dawn. The bright brilliance of bolts of lightning electrifying the air during the season's first thunderstorm. The splendor, joy, and wonder of watching the fall foliage transition. Some of the most amazing shows are written, directed, and orchestrated by nature. Better yet, we often have a front row seat and backstage pass, and they're free! It's been well substantiated that experiencing nature is good for our mental and physical health. Some of the best things in life are those that bring gratitude, joy, and love. When we selflessly express love to ourselves and to those around us whom we wish to extend it to, we serve a purpose larger than ourselves. Everyone has innate gifts that define who they are. Such gifts can be as subtle as a smile to as sophisticated as singing a stanza of music. Exploring and sharing our gifts embodies what it is and means to be human.

Treat the planet as your home. We are not separate from each other, the planet, or the universe, for that matter. Although our homes provide shelter and sanctuary from the elements and comfort us, our original home remains the planet. No matter where we live, we are all occupying the same big rock. How we respect ourselves, our homes, and our neighbors says a lot about how we respect the planet. Mahatma Gandhi is quoted as saying, "*We but mirror the world. All the tendencies present in the outer world are to be found in the world of our body. If we could change ourselves, the tendencies in the world would also change. As a man changes his own nature, so does the attitude of the world change towards him. This is the divine mystery supreme. A wonderful thing it is and the source of our happiness. We need not wait to see what others do.*" The quote that many people have attributed to Gandhi, "be the change you wish to see in the world," is a derivation[112] from his original statement. Whichever quote you choose as your favorite, the fact remains that modeling leadership changes behavior. When we align our principles with our behaviors, even those that are simple and practical, we have the power to influence others, shift mindsets, and change our world.

The following Profile in Pragmatism examines how a dear friend and colleague, Rajiv Ramchandra, chose to serve his purpose in the ideation and development of the ReCREATe India Research Foundation (ReCREATe). Established just before Covid-19 was a roaring pandemic, ReCREATe serves to create a thriving remanufacturing industry in India.

Profiles in Pragmatism
Pursuing Prosperity through a Regenerative Industrial Economy — Rajiv Ramchandra, Founder and Chair, Board of Directors of Recreate India Research Foundaton (Re:CREATe)

Rajiv Ramchandra is the Founder and Chair to the Board of Directors of the Recreate India Research Foundation (Re:CREATe). The vision[113] of ReCREATe is "to be co-creators of a thriving remanufacturing industry in India," and its mission, "at Re:CREATe, we are deeply committed to the vision of noble environmental stewardship. We honour this commitment by being advocates for, and providing tools, resources, guidance, and inspiration to the makers, doers, and designers, empowering them to develop and provide remanufactured products and related services that are in harmony with the principles of environmental stewardship."

I've had the pleasure of not only knowing but also working with Rajiv and a friend and colleague for over fifteen years. We first met when we worked at Rochester Institute of Technology (RIT). Rajiv worked for the New York State Pollution Prevention Institute (NYSP2I), which is housed at RIT, while I worked for RIT's Center for Integrated Manufacturing Studies (CIMS) and Clean Energy Incubator (CEI). We worked under the same administrative leadership, and there was camaraderie and synergy between our respective organizations. Our shared experience at RIT was also firmly grounded in foundational applied research and technology development centered on remanufacturing and remanufacturing technologies. Dr. Nabil Nasr, world-renowned for his leadership and work in pollution prevention, remanufacturing, circularity, and sustainability, oversaw the respective institutes and centers that Rajiv and I worked for. Naturally, we were exposed to all things remanufacturing.

I fondly remember many breakfasts and lunches with Rajiv. The Highland Diner in Rochester was a regular favorite, although we frequented many local eateries. Like all great cities, Rochester has its fair share of iconic eateries. Our conversations were always quite natural. Often, we'd discuss work, but usually we found ourselves immersed in conversations that transcended life and spirituality. There is something special about those friends who allow you to be yourself, open, vulnerable, and honest with yourself.

Since our early days of getting to know each other, Rajiv and I both departed from RIT to explore and expand career opportunities, respectively. In Rajiv's case, he travelled back to his home in Mumbai, India, for a few years before moving to Calgary. We've kept in touch over the years, at least monthly, but often more frequently. During those foggy months of COVID, Rajiv had a FaceTime call with my two sons and walked around Mumbai, showing them his neighborhood and a few historic landmarks. It was a fun way for my sons to learn a little about India, and through the eyes of someone who was rediscovering the people and land he had originated from. Rajiv was kind to answer a few innocent questions from my sons as he provided a guided tour of the city near the shores of the Arabian Sea.

Curious, wise, intelligent, kind, generous, introspective, and funny — Rajiv is many things, and by virtue of his characteristic traits, he also brings out the best in others. When Rajiv called me years ago about his intention to launch Re:CREATe, I was ecstatic. He had been back in India working for a sustainability consultancy and had inadvertently stumbled upon a blatant gap in their regional economy. Rajiv discovered that remanufacturing and circularity were largely vacant from the broader economy outside of a handful of companies working in that space.

Naturally inquisitive, Rajiv began digging deeper and held conversations with corporate, government, not-for-profit, research, and academic organizations. Quickly, he discerned that there was an opportunity to aid in the advancement of remanufacturing and a circular economy in India. Rajiv created a remanufacturing roadmap that was reinforced by a Theory of Change model and then proactively sought out the input from key leaders throughout India, which resulted in multiple iterations of the roadmap. Eventually, Rajiv's efforts led to the derivation of five[114] essential needs and pillars of change, which included:

- Collaboration
- Research
- Education, Training, and Incubation
- Advocacy
- Technology

The identification and prioritization of these pillars provided a foundation that would lead to the launch of a new social enterprise, the Recreate India Research Foundation. Although he had limited resources, Rajiv chose to listen to and explore his intuition. Rajiv's desire to leverage his prior knowledge and experience while exercising his strong facilitation and business development skills led to a deep insight regarding the existing institutional, economic, and social barriers that have challenged the evolution of India's remanufacturing industry and regenerative industrial economy.

With fortitude and conviction, Rajiv began to establish a global network of supporters. Rajiv put the Theory of Change model into practice. ReCREATe began engaging stakeholders, developing research white papers, participating in industry events, facilitating peer-to-peer workshops and panel discussions, advocating for policy, and much more. With persistence and perseverance, the purpose and goals of ReCREATe become more known to leaders in industry, government, and academia.

In recognition of his contributions, in 2020 Rajiv Ramchandra and ReCREATe earned the distinction of *"Best Reman Ambassador"* from Rematec, the world's leading platform for remanufacturing and the only independent trade brand dedicated to automotive remanufacturing. Rematec's award announcement[115] stated, *"With Re:CREATe's vision to be co-creators of a thriving remanufacturing industry in India, the organisation has made a sound impact on the jury with their commitment to being advocates for, and providing tools, resources, guidance and inspiration to the makers, doers, and designers, empowering them to develop and provide remanufactured products and related services that are in harmony with the principles of environmental stewardship."*

Planet pragmatism is about examining our intuition, challenging convention, and pursuing a more prosperous future. Rajiv Ramchandra's manifestation of ReCREATe personifies the faith, wisdom, and courage that is so often the formula when charting new paths.

Rethink your relationship with ownership. Ownership is a choice, privilege, and responsibility. If you are going to own something, consider the short, mid, and long-term implications of your decision. Ownership brings a responsibility for maintaining, and even upgrading and improving, your assets over time. Alternatively, non-ownership models can deliver the utilitarian benefits of what you may need (i.e., housing, mobility, office space) without all the responsibilities that an owner has. Owners who have and maintain deep principles for sustainability demonstrate a level of stewardship that is greater than those who simply want to own something just to own it or exclusively for financial gain. This is not to say that financial gain from ownership is a bad thing. Instead, ownership that is rooted in pragmatic and responsible stewardship leads to value creation that should and can be rewarded financially. The difference comes down to principled leadership and values the owner places on their stewardship of an asset.

Explore new economy shifts, including those associated with the Seven Ds and the shift from consumer to prosumer. There is no dispute that a global shift is underway, marked by a changing economy, demographics, social expectations and needs, and a redefinition of what prosperity really looks like. As described previously in this book, the Seven Ds (now eight! meta dimensions) of sustainability are driving a sea change in thought and action toward planet pragmatism as a new path to achieve prosperity. Prior generations developed and lived with economic models, typically linear and sequential in their design and orientation to deliver resources to society at scale. These models were prided on their replicability, predictability, and scalability, particularly about providing profit to business owners and investors. The American Dream and definition of prosperity for much of the last century were interwoven with this consumption-based economic model of linear growth. These models did not account for environmental or social externalities that should have been factored into the economic equation, or the probability that certain ecological services or resources would become constrained and limited. These models made confident and unassuming predictions that growth was infinite and that the world's resources and capacity to absorb our waste would always be sufficient to accommodate our linear progression. Well, our predecessors were sorely wrong in their assumptions, or they chose, perhaps ignorantly or selfishly, to dismiss the realities of unfettered growth within a finite system of resources. As we know, the human population grew quite a bit over the past century, as did our desire for more energy, water, food, and other resources, and that has delivered a technically sophisticated, yet ethically immature society. As more people in developing countries enjoyed the benefits of a vast consumer-centric economy, the externalities of this ubiquitous culture, including environmental pollution and social ills, began to rear their ugly heads. Rivers were polluted and ablaze, nature was dying, people were falling ill from diseases. In short, our linear models of production and consumption became dated and deficient in serving the evolving needs of society. Fast forward to today, and we continue to hold onto our past economic structures while we desperately attempt to shift our thinking and behavior into a new paradigm and more complete economic model that is more inclusive, sustainable, and attainable for society. In my children's lifetime, less than two decades, there has been a devolution from the consumer-based economy toward a "prosumer" economy where the consumer can be a producer and consumer. The meta-dimensions including decentralization, digitization, decarbonization, and democratization each align to stimulate the prosumer economy, with electric vehicle's being an example where consumers will soon have the potential to consume electricity form the grid to power their vehicle and also have the prosumer option to push power back to their home (for in-house electric utilization) or the power grid (power-grid utilization) from their vehicle (i.e., vehicle-to-grid, V2G). The shift from consumer to prosumer to highly integrated circular and restorative economies is underway. This shift will take time as it challenges convention and everything that the current and past economy thought was true about resource economics.

Investigate and question your relationship with anything and everything, aka, stuff. Think about what is essential in your daily life versus things that are thus nice to have. Most every product creates waste. Explore your waste, figuratively and literally. Anytime you see waste associated with the financial premium you are paying to account for someone else's lack of creativity, foresight, and innovation to provide you with a better product. Anytime you experience waste, you are choosing to become the responsible party of that inefficient design, be it a coffee cup, a water bottle, a clothing garment, a piece of furniture, or a new automobile. When you begin exploring the waste in your life, you will see inefficiencies that can be overcome through better planning and with solutions that lead to healthier choices for you, your family, and the planet. Waiting for business to solve all the waste and inefficiencies in our consumptive society only perseverates the status-quo consumer mindset and limits your capacity to be a planet pragmatist, that is a straightforward, clearheaded steward of living better, now. If you don't need something, don't buy it. If you can give something away to someone else in need, do so. Declutter your home, your office, and a decluttering of your thoughts and mind will follow. Let the wave of simplicity wash over your life. Living with less can result in living with much more when you let your principles and values guide your consumptive decisions.

Exercise agency. Regain control of your everything. When I say regain control, I don't mean this in the egoic God-like domination that humans have brought upon the planet to date. Our grand attempts to tame nature have historically been met with the true force majeure of nature and God and have always skewed in their favor. Yet, we simpletons are slow learners, so we try and try again. Instead of universal dominance, when I say control of *your* everything, what I am referring to is those attributes and elements that you can, reasonably, control for, meaning your time, your mind, your finances, your relationships, your education and professional development, your career, your family, your community, your health, your life — your everything! If you haven't noticed, or if you've been asleep at the wheel or slow to rise from the couch and underneath the blanket shielding you from the trauma-induced years of COVID, demagogue politicians, and social unrest — well, wake-up, rise, have a sip of coffee, and put phone back in its proverbial holster. Your time and attention have been hijacked by one of the most iconic and celebrated products ever made, the smartphone. In the past few years, our smartphones have become weaponized, for and against us. Our data and our privacy are being stolen, exposed, and under attack. Like millions of knighted warriors heading to battle each day, we carry mini shields of armor and swords (i.e., our phones) that we use to slay enemies and protect us from unassuming attackers. As the adage says, we may be winning the fight, but we are losing the war. The fight may be a daily exchange over text, sifting through chatter on social media, or posting a new selfie. The war we are losing, however, is the mass distraction, obfuscation, and deterioration of our lives that is taking place, in front of our eyes, at the hands of our own doing and our willing minds. Without restraint or reflection, our continued immersion into selfish pursuits and gratification via our "interconnected society" is yielding a greater rise in mental illness and people feeling truly alone than at any other time in history. We're connected, but our lives and relations are strained. It's time to take back who we are so that we can become who we need to be.

Don't overthink. Simple truths don't require complex solutions. Overthinking can lead to delayed action. Action, even if imperfect, provides feedback that is essential to ongoing improvement. The planet is incredibly complex, yet it is elegant, simple, and beautiful. Every sustainability challenge and opportunity begins and ends with us, as originators, stewards, and solvers. We have the knowledge, know-how, and tools to be preventive, predictive, and proactive in our posture toward how we live in greater harmony with each other and the planet. We must get out of our own way and limit the desire to overcomplicate our world. The sustainability (energy, economic, environmental, social, healthcare, education, transportation, housing, etc.) we see in the world today is typically symptomatic of a deeper entrenchment of root causes. These root causes can be "artificial" in that we (society) generally tolerate and allow them to continue to exist. Rethinking and rebuilding a better future does not require grandiose or sophisticated plans. By questioning and restructuring the prior plans and existing rules, systems, and capital that comprise the modus operandi of the classic institutions that govern society, we can rectify the root causes that perpetuate modern injustices and unsustainable behaviors. New technology, businesses, and ways of thinking are challenging older paradigms each day. Blockchain, for example, is deconstructing the centralization of finance, energy, real estate, and more by providing a distributed ledger supporting a more complete and trust-based accounting for data transactions.

Do not let the pursuit of perfection get in the way of achieving something great. Sustainability is not a binary (achieved or not achieved) outcome. Sustainability and planet pragmatism are not about being perfect. It's about having grit, that is, the courage, resolve, and strength of character to take on any challenge, pull through any moment, and come through the other side, stronger and better. Humanity demonstrated grit through COVID. In the face of global fear and uncertainty, humans displayed deep compassion, courage, and an innate desire to conquer COVID. Whether it was one-on-one compassion, communities working together to administer test kits and vaccines, or the groundswell of entrepreneurial spirit that chose to innovate and then build back better together, society got through COVID, in part, through some incredible displays of grit. We may not be perfect in cleaning up our room (i.e., reducing greenhouse gas emissions, cleaning up contaminated lands, drastically reducing plastic waste from waterways, ensuring every child breathes clean air and has access to clean water), however, that does not give us permission to live like pigs in a sty. Our objective is to continuously improve upon the human condition. Enhancing our quality of life requires that we value, protect, restore, and respect the planet. Our fate and that of the planet are one and the same. Thus, for us to have a high quality of life, we need a clean, but not necessarily perfect, room.

Take stock of, and celebrate, the great things around you, now. The future is now. Every decision we make here at this moment affects the future. Humans tend to place a great deal of emotional emphasis and personal resources (including time, thought, and brainpower) on the future state of our lives. We are thinking about our next meal, perhaps a nice dinner with our partner. We envision ourselves at that favorite vacation spot, taking some much-needed rest and relaxation, refueling our soul. We chart out a path to achieve that professional certificate to ensure we remain competitive and can continue to grow within our chosen career. We proudly foresee the moment our child receives their high school diploma and eagerly await what's next for their blossoming life. We fret whether we are saving enough for retirement and whether we are doing the right things now by way of diet and exercise to ensure we are healthy in the future. All of these represent future goals or objectives that we focus on, as we want to see their positive manifestation. Our quality of life and happiness are shaped by our decisions over time. There is a delayed gratification to some of our decisions and the impact we see, for example, losing that extra fifteen to twenty pounds through good nutrition and exercise. Our quality of life and happiness are predicated on our state of mind, our decisions and actions in the moment, and our ability to reconcile our mind's conscious and subconscious goals with our wants and needs. Quality of life and happiness are not a future state of being. We can attain and achieve a higher quality of life and sense of well-being at any moment by being present and expressing gratitude for the great things that exist within our lives and all around us. Our subconscious mind may want us to believe that we'll only be happy if we're on a yacht sailing in the Mediterranean. Our conscious and critical mind may say, we will never save enough money or have enough vacation time to make a trip to the Mediterranean. We must be active stewards of our intrusive and negative thoughts; otherwise, our minds will default into overthinking, perhaps to the point of indecision, where we fail to establish a goal or put a plan of action into place. As a result, we sit idle, spin, and stew in our overthought and judgment of our "lack thereof," allowing ourselves to feel unsettled, unsuccessful, and unhappy. We have the power to limit our own minds' control of our negative thoughts and the onslaught of emotion that comes along with this phenomenon. Self-awareness, being present, and finding joy in every moment are all skills and tools that can lead to greater happiness and quality of life. Listen to the dialogue that's happening in your mind. Often, the dialogue is not focused on the present, but something that happened in the past, or something you're fixated on in the future. The past is the past; that moment, whatever it is or was, is gone forever. So, let it go. The future is robbing you of living your best life in the present moment. If you live your life where every decision is toward "a better you" that exists in the future, you will never truly enjoy the current moment you are living in. You will delay gratitude and gratification forever. This is why it is critical to take stock of all the great things around you, so that you can ground yourself to be present and feel the energy and life that surrounds and uplifts you right here and now. This also aligns with the principle of getting back to nature, back to basics. As we all know, there are too many issues that hit us squarely between the eyes every waking moment of every single day. Take a moment to hit pause and let the Sun warm the skin, the breeze flow through your hair, your hand glide through the water. Our senses were built to sense the stimuli of our incredible planet, here and now. When our senses are activated, so too are our minds.

Points on Pragmatism

- *Wisdom is useless unless we put it to use. Essentially, we need to learn from the past. And we need to stop doing stupid things on purpose.*
- *Let's show some humility and restraint, especially when it comes to advanced technology. Just because we can, doesn't mean we should.*
- *Nobody wins in war. We must end unnecessary conflict and acts of aggression that cause undue harm to people and planet.*
- *Disagreements can drive creativity.*
- *Scarcity and necessity beget innovation; but remember, complex problems don't necessarily require complex solutions.*
- *Prepare for the future but live for now.*
- *Redefine and rebuild communities that foster quality of life and "real" community.*
- *Stop blaming others and start taking responsibility, accountability, and ownership over our pursuit of prosperity.*
- *Redefine wealth and prosperity on your terms. To live simply is to live wealthily.*
- *Raise the floor and the ceiling on sustainability education and upskilling.*

- *Go beyond conscious consumerism, and ask yourself what you really need to be happy, content, fulfilled in your life.*
- *Get back to nature, get back to basics.*
- *Treat the planet as your home.*
- *Rethink your relationship with ownership.*
- *Explore new economy shifts, including those associated with the "Seven Ds" and the shift from consumer to prosumer.*
- *Investigate and question your relationship with anything and everything, aka, stuff.*
- *Have agency. Regain control of your everything.*
- *Don't overthink. Simple truths don't require complex solutions.*
- *Do not let the pursuit of perfection get in the way of achieving something great.*
- *Take stock of, and celebrate, the great things around you, now.*

14

FIVE PILLARS OF PLANET PRAGMATISM

This book was written to explore the limitations and potential of sustainability as a foundation for pursuing prosperity. The pursuit of the "S-word," sustainability, should be compelling enough. Yet the politicization and weaponization of terms like sustainability have created confusion and devalued their purpose among some members of society. The pursuit of greater prosperity remains a central tenet toward the ongoing evolution of humanity. Our collective history shows that humans go through cyclical transitions of suffering, growth, innovation, abundance, and decline. The fate of humanity has become intrinsically tied to how well we learn from the past and apply ancient knowledge and innate wisdom to the pursuit of prosperity, that is, common sense for the common good, through planet pragmatism.

Planet Earth is and shall remain our home, at least until we are rendered dust within the cosmos, or discover the economic, equitable, ethical means to uplift and transplant ourselves off Earth and then travel to and terraform another planet. The fate of the Earth and that of humans is the greatest story we know. Humans are the lead characters, tragically flawed, beautifully brilliant, and eternally optimistic. Our story is intertwined with nature, the Universe, and each other in ways that we have not been able to fully comprehend. Shared human wisdom and experience tell us that life is precious, and it can certainly be fleeting. Our time is functionally limited, and yet, we have the option to maximize its impact in all that we are and do. Planet pragmatism is the recognition that we are all connected with each other, with nature, with the Universe, and with the metaphysical and spiritual worlds. We have the necessary skills, tools, and resources available to us right now that can enrich our lives and bring about greater prosperity for all people of the world.

The definition of prosperity differs among individuals, cultures, and nation-states. However, common threads that intertwine prosperity include having the freedom to pursue one's fate, enhancing the quality of life for all people, being able to have equitable access to natural resources, and knowing that dignity and well-being for all living things are a foundation for survival. This chapter closes the book by laying out five pillars of planet pragmatism. These pillars are important to the ongoing survivability and sustainability of humans and life on Earth.

Like all frameworks, the content, ideology, principles, and ideas outlined within this book are entirely open for critique and debate, subject to evolution, and intended by the author to be challenged and refined. This book came from a place of personal introspection, experience, learning, and creative expression. In that, it has always been my intent and hope as an author to openly explore and express new ideas, challenge convention, and elevate logic and wisdom as cornerstones for personal growth and social progress.

As you consume the final pages, please know that I am deeply grateful for your time and attention. I cherish the fact that although we are living in divisive times, we have agency to take control of our lives and reframe our future for the better. At any moment, we can positively impact our lives and the lives of others. We can learn to better manage our fear, anticipate and adapt to change, redeploy resources, restore nature and nurture our creative souls, and rediscover and reinvigorate the power and delight of working together for the common good. Simple and elegant, planet pragmatism harnesses the ancient wisdom that is encoded in all living things and unlocks the knowledge and understanding that our prosperity is, as it always has been, universally interconnected.

Pillar I. Mind mastery, rather than our mastery over nature, may be one of the most compelling and pragmatic remedies toward creating a more sustainable planet. Wisdom is revealed in the calm quiet, when our minds connect to a higher consciousness. To accomplish this, we must manage the rampant and unproductive thoughts within our minds. Mastery of one's mind is a cornerstone of individual compassion, leading to a more benevolent society and greater prosperity. Too often, we fall short in our individual and collective capacity to willingly exercise grace, dignity, and temperance.

Instead of trying to tame nature and control all living things, including the behaviors of our fellow neighbors, what if we chose to focus our time and attention on mastering our own personal thoughts, beliefs, actions, and reactions? Why is it that we allocate so much energy and time to trying to control nature, people, and the vast world around us? What if we tapped into our innate wisdom and acknowledged that all living things are intrinsically linked, and therefore we have an obligation to live with nature, not against it?

Anger, resentment, jealousy, envy, fear, apathy, anxiety, discontent, and a host of other emotions are manifestations of our mind's restless reasoning with itself against external stimuli. Objective, clear-headed reasoning and logical thought can provide an antidote to a wandering, mischievous, and wayward mind. If you've ever had a runaway thought that made your mind spiral in a million directions, you know exactly what I'm speaking of here.

Perhaps after witnessing your partner smile at one of their colleagues, you had a misguided thought of them being unfaithful to you. Or perhaps you found yourself creating false scenarios of all the bad things that may be occurring to your child because they didn't answer the phone right away when you called to check in on them. Or perhaps your insecurity has exploded due to a lackluster recent performance on a job task that now has you overthinking your relationship with your boss, creating unnecessary anxiety and sleepless nights. Our mind is continuously assessing, evaluating, and judging all aspects of our daily life and existence. Our minds operate in complex ways. In fact, our brains are the fastest, most dynamic, and intuitive processors known to humans. This is why leading scientists and technologists modeled neural networks used within artificial intelligence (AI) from the human brain.

The brain is a blueprint to higher intelligence, an intelligence that we've only begun to understand, let alone wield the power of. So put down and step away from your iPhone, iPad, or Google device. Your brain is more powerful and has greater capacity than any digital device on the planet. But, unlike our favorite gadgets, none of us came with an instruction manual, and we don't have access to a call center for "brain" support. Of course, there are professional services that people can call upon in times of emotional, health and safety, psychological, personal security, and other needs. The call center I'm referring to here is the one that can unlock our higher intelligence, enabling us to achieve prosperity in a dignified way, by respecting and uplifting all living things.

How can we crack the code to unlock the full potential of our meta-physical, cognitive, emotional, and other intelligences? We have the intellect, that is, the brain power, to figure out a sustainable way of life. The question and challenge are, however, how do we harness our enormous intellectual capacity to create meaningful, lasting, sustainable change? Underlying any enormous computer is the code, or instructions on how to compute. The neurons within our brains contain and process infinite lines of code, many of which we haven't even revealed yet. However, based upon centuries of observed human behavior, we tend to rely very heavily upon a few select lines of frequently accessed code, particularly those that pertain to our ego, emotional response, sense of self, and survival. Humans are, after all, hard-wired for survival — there is a code within us, running nonstop, always attempting to protect our physical, emotional, and spiritual being.

For example, when our mind runs rampant, our logical control center begins to shut down, giving rise to our fight or flight responses used to protect ourselves from external pain that our mind has perceived, or even fabricated. We play out mental scenarios of what could happen as a measure of preparedness, so that should one of those scenarios become true, we will have had time to consider (and prepare for) the impact, consequences, and pain. Have you ever thought about the death of a loved one and the sequence of events that may follow? A morbid thought, but it is also part of being human and working with a big brain that stratifies between conscious and unconscious thoughts and feelings all day long.

Consider the positive, negative, or neutral self-talk inside your head at any given time. What's it saying? What's it projecting? What's it asking? What is your self-judgment mind trying to protect you from? Many books have been penned, and many "experts" have elevated their points of view on human psychology, personal growth and development, and how one can better control one's conscious and subconscious thoughts, actions, and behavior. Yet amid all the content and pontification on how we can control our thoughts and our futures, humanity falls way short of advancing our intellect for the common good.

None of us is perfect in the journey toward practicing planet pragmatism. There is always something to learn and relearn as we travel through the arc of one's life. In the sidebar below, I share a personal story about how foundational principles of pragmatism, including self-restraint and awareness, accountability, and emotional intelligence, are practiced and relearned over time.

The Leaning Tower of "Pizza": How the Principles of Pragmatism are Practiced and Relearned Over Time

Recently, I found myself getting schooled. Who was the teacher, you might ask? Well, that was my son. Any parent who has had this kind of experience knows how humbling it can feel to have your child point out an obvious deficiency in your psyche and behavior. When this moment happened, I instantaneously flashed back to my 16-year-old self and remembered similar interactions with my father. I felt my son's frustration and took away a valuable lesson.

Here's what shook down. It occurred during a post-holiday evening, but for all purposes, a normal dinner night. I had just begun to get some good eats ready for the family. My son, hungry yet patient, took a seat at the kitchen table. He had his phone in hand and sat scrolling to pass the time. He had already gotten himself a glass of water. He's always helpful in setting the table, pouring beverages, and laying out any condiments if needed.

As an important backdrop to this story, my son has severe food allergies. Our family shares some of his food. And then there are foods that we make uniquely for him because he enjoys them. As such, we have a lot of separate storage containers in our refrigerator. In support of managing our son's food allergies and diet, over the years, we became highly proficient in safe food preparation, cooking, and storage. With the precision of a surgeon, we mastered the art of making healthy, delicious, and safe foods for our son. Now seventeen years old, our son watched us over the years and has learned a great deal. He now helps a great deal in the kitchen, making his own food and offering suggestions along the way. This backstory is important, as anyone who has children with food allergies knows, food management is both a tactical daily practice and an ongoing emotional (more for the parents) journey. For seventeen years, I've prided myself on the military discipline of food organization. That is where this story begins.

It was an evening in early January 2025. Like most evenings, I was dancing around the kitchen (literally and figuratively), as I pulled select items out of the refrigerator and laid them out on the counter. Being post-holiday, the refrigerator was brimming with leftovers and had more containers than usual. I shuffled and reshuffled the food container deck several times, placing what I needed on the counter and moving unwanted items to the back of the refrigerator. Where else would they go, right?

I began plating dinner for my son. I realized there was something missing. So, I then turned back to the refrigerator, opened the door, and began shifting containers, searching for the missing delight. But then, as if the physics of reality were altered, the leaning tower of food container "pizza" (pun intended) within the refrigerator suddenly tipped. Container after container seemed to slide out of the fridge, crashing onto the kitchen floor. Food exploded everywhere.

An explicative may have ignited and spewed from the frontal lobe of my brain in tandem with the food that splattered the floor and surrounding cupboards. In my haste, I created waste and certainly a mess that needed to be picked up. The food had not been on the floor a few seconds, and my son peered up from the dinner table, and with a not-so-subtle sarcastic tone said, "Why don't you just label the containers so that they can be identified better in the fridge so that they don't fall out? If you did, maybe that wouldn't happen."

Oof, my son's immediate feedback was honest, and it hurt. Upon hearing my son's response, my first instinct was to refute his claim and fire back. But I restrained myself. After all, the overflowing refrigerator, the leaning tower of containers, the mess on the floor — all of these were my doing, certainly not his. Any negative reaction I might have would be tied to my lack of emotional intelligence and control. When I separated his snarky tone from what he was saying, I could not dispute his logic. The refrigerator was unorganized. There are simple ways to label and separate his food from others to make it accessible, convenient, and sanitary.

Deep down, I knew all of this, as I was disciplined at it for years. For so long, I had labeled foods and generally kept a tidy refrigerator. But my discipline had waned, and I was now wiping up the kitchen floor while choking on a piece of humble pie. I told my son that he was right. Now, this event only resulted in some spilled food and a small mess to pick up. Thankfully, there were no food allergy concerns over food cross-contamination or mislabeled containers, and so on. The event was, however, a teaching moment for me.

Occasionally, we may get lackadaisical and complacent in how we manage important elements of our lives. But it is in these exact moments when the physics of reality can slap the containers right out of your hands and wake you up from your fog. Self-control, emotional intelligence, gratitude, owning up to one's mistakes, accountability — these are all traits that we learn and relearn over time. They are also foundational to the skills of pragmatic leaders and planet pragmatism.

Being more mindful of one's mind is nothing new. Ancient Greeks and Romans practiced stoicism, a Hellenistic philosophy, which focused on four virtues: wisdom, courage, temperance or moderation, and justice. In addition, Stoics believed that a life well spent was one that also lived in accordance with nature. Stoics and their philosophy believed that a person's values were best characterized by how they behaved, not by what they said. The Stoics believed everything was rooted in nature, and therefore, to live a good (prosperous) life, the rules of the natural world needed to be understood, valued, and respected.

From my early childhood, I fondly remember my maternal grandfather shouting out, "Marcus Aurelius," when he would see me. At the time, I had no idea who Marcus Aurelius was. But the way my grandfather said, "There he is, Marcus Aurelius," with joy when I entered the room. My grandfather's gesture always immediately provided me with a sense of pride. I could sense that Marcus Aurelius must have been an important man. Years later, I would learn about Marcus

Aurelius[116], Roman emperor from 161 to 180, and Stoic philosopher whose writings, Meditations, offer a glimpse into how he internalized and reflected upon the four virtues. We all know that a great deal can be learned from history. Our current culture, society, and individual DNA, core beliefs and values, and minds have all been formatively shaped by history.

Marcus Aurelius' *Meditations* and the Stoic philosophy are as relevant today as they were thousands of years ago. That is the elegant nature of having virtue as an operating system, and clearly defined principles as the filter by which to guide the code running within our brains. When we lead lives of virtue, the operating system remains evergreen, allowing us to create and modify principles that can lead us through each subsequent era of human evolution with a sense of ingenuity, integrity, and resolve. Pragmatism plays a significant role in maintaining our virtues over time. Practical action, grounded in ancient virtues of wisdom, courage, temperance, and justice, can ensure a focused and clear mind for individuals — yielding greater cohesion toward the common good of humanity.

Pillar II. Learn objectively from the past and lean intentionally into the future. Achieving prosperity requires us to reframe how we define and determine success. Time is our most precious asset, more so than money or technology. Respecting time and using time wisely is a key pillar of Planet Pragmatism.

A friend once asked me my definition of success. Before I provided an answer, I took a moment to reflect and truly consider my response. My definition of success has evolved over time. As a teenager, I looked at material possessions like exotic cars and well-manicured mansions as illustrative of success. In my twenties and thirties, reputation, social status, and power became symbols of success. Then, in my forties, family and relationships superseded all else in my self-evaluation of what success looks like. And now, on the brink of living for half a century, the quotient of time has become a precondition of success. In many ways, success is not a predetermined one-size-fits-all metric.

Over time, my views on success have evolved in step with my life's journey as a student, as an employee, as an author and entrepreneur, as a brother and son, as a husband, as a father, and as a human. I don't believe there is a singular best answer to the question, "What is success?" The answer truthfully resides within each person's soul. Success equates to several components that come together to enrich one's life, and more importantly, to empower and enrich the lives of others.

This said, I don't feel qualified to pontificate on how another person should define success, let alone how another person should live their life. The wisdom and insight of others, from ancient philosophers to grandparents who have passed on, to newfound friends, all contain morsels of truth when it comes to evaluating success. As I grow older, I find myself peering back to the writings and philosophy of Stoics, including Marcus Aurelius, for wisdom. Language can be limiting, and the idea of success in and of itself reduces the full intellect and potential humans have. The Stoics, for example, were keen on pursuing a "life well-lived." The idea of pursuing and measuring a life well-lived feels more meaningful than simply defining success.

At this point in life, exploring the deeper intention behind life's purpose, including one's individualistic and interconnected experiences, resonates a higher frequency within me rather than trying to define success by a singular metric like financial wealth, social status, or personal accolades. So, what is life, well lived? What are the characteristics of having spent time wisely, efficiently, productively, and virtuously? There is nuance to how we choose to communicate our perspectives on prosperity. In this emerging era of AI, we should heed caution to the automation of language. AI models and tools have their utility, but they can never serve as proxy to what we hold in our hearts and minds. When it comes down to defining the virtues and principles that should envelop humanity's pursuit of a more prosperous future, humans need to intimately be "in the loop," so to speak. Our voices, intentions, and hearts can never be subverted by AI.

When my spouse and I first had our children, so many families and friends, particularly those that already had children, would say, "Enjoy being a parent, it will go by fast." We always politely acknowledged and accepted the unrequested advice, nodding and smiling kindly. As any new parent knows, those first few sleepless nights and exhausting weeks and months of caring for a newborn baby don't exactly align with the swift passage of time. In fact, just the opposite. The seconds of the clock tick-tock in a mocking manner when trying to soothe a colicky baby. Time seems to stand still when one is experiencing discomfort.

As the tiring and arduous days and nights of nursing babies give way to the spunk of toddlers and rambunctiousness of early childhood, time seemingly begins to move at an ever-quickening pace. By the time the children are adolescents and young adults, you find yourself gasping and grasping for air, trying to catch your breath and slow everything down. It's at this point when you remove the freshly brewed mocha latte from your lips, put the cup down, and break out in a cold sweat realizing that one, your stomach can't handle milk products or caffeine anymore, and two, those damn family and friends were dead right sixteen years ago. How time flies!

The human psyche, in the absence of strong virtues and principles, is caught up in its own contradictory state of being. On one hand, humans want to achieve greater prosperity; on another, we are selfish with our time and self-destructive with one another. We inherently know that time is precious, yet many of us continue to squander it. The bookends of our life, when we are born and when we die, represent the space and time by which we are alive. When we bind our existence, we tend to fixate on living in the moment, not just living to survive. Our hard-wired sense of survival sometimes gets in the way of us actually *living,* that is, taking full stock of and optimizing our time, making the most of our precious lives, between the bookends.

We must not forget that survival is an essential requirement for living. In recent years, I've begun to think of success as having the freedom to operate and do good things. In this broad interpretation of success, I value time as one of the most critical elements. Time provides us with the utility to explore the freedom to operate. If one chooses to pursue an educational degree, learn a skill or trade, immerse oneself in music or arts, take that long-awaited family trip, or provide a much-needed service to others, time is the underlying factor that is either limited or abundant.

External measures of how one chooses to define success (i.e., money, power, influence, material possession, and even family, friends, relationships) are each enabled by the quotient of time. In this way, my personal definition of wealth is multifaceted, but central to it is having an abundance of time — time to reflect, time to write, time to learn new things, time to soak up every ounce of my boy's maturation into adulthood, time to take a midday hiatus with my lovely spouse.

Time, if optimized, can aid in the derivation of any of these and other indicators of personal gratification and/or personifications of success. Therefore, the freedom to operate, or choice to freely explore, is predicated on having the necessary allocation of time in your life to pursue your desired interests. To me, that is living richly and illustrative of wealth. Not everyone has an abundance of time. Some might say, *well, if I had financial abundance, then I could have time abundance.* I have personally spent half a century locked up in that frame of mind and cycle, and the one thing I am certain of is that money cannot buy back time. If we choose to value and use it wisely, our lives can be enriched by time.

When we are younger in age, our brain interprets the time available in our lives as abundant. As we age, and as the passage of time yields life experiences and outcomes, the brain begins to recalibrate its understanding of time — determining that we are time-constrained. My singular experience with time is not unique. The human experience and our individual journey through life and collective story through the ages have been ones that have tried to reason with time. Time has no ability to reason with mortal beings, however. Fundamentally, it is a human construct of measurement to provide context to what happens between our life's bookends.

Metaphysically, time is more dynamic than what we experience between birth and death. Metaphysical time encompasses an infinite expansion that includes our origin story, our current story, and our future story. Given the dynamic nature of time, I personally believe we should not fear the unknown, as perhaps the bookends are not really bookends at all; rather, they represent events across the sands of time. This said, our mind craves and adores certainty; thus, there is nothing inherently wrong about valuing our physical representation of time between the bookends. The trick is figuring out how best to make use of this beautiful thing we call time.

Pillar III. Temperance and moderation are ancient philosophies that remain pillars of Planet Pragmatism. Measuring worth in material possessions is false and fleeting. When lived with virtue and in a principled way, life delivers deeper riches. You, me, and we can live prosperously with very little.

How do you define success? What is your definition of prosperity? What constitutes a "life well spent?" How do you want to spend your precious time between the bookends? As previously mentioned, my intention is not to tell you or anyone how they should live their life. That is 1000% up to you and them. What I am attempting to do, however, is have you hit pause, interrupt your schedule, and disrupt your typical thought patterns, and self-evaluate your perspective on prosperity, joy and happiness, success and freedom. As you do so, think about your real relationship with material items. Ask yourself, do your relationships, material possessions, and how you utilize your time align with your overarching virtues and values? Is your life aligned with your definition of prosperity? If you feel that you are not successful in some way, can you self-identify and define why you feel this way? What is it that you feel you are lacking?

Material possessions are simply an extension and abstraction of how we choose to use our time and an edification of the value we place on measuring our time. When we consider material possessions, we should do so with a certain degree of ambivalence. Most of us, I would argue, fundamentally, are not concerned with the accumulation of material possessions as the primary indicator of one's life or success. That is not to say that we do not get consumed by the temptation and allure of material items. This is not entirely our fault.

Our global society and economy have been constructed to intentionally drive capitalistic pursuits and consumer behaviors. Success and prosperity in our modern era require us to be careful stewards of our minds so that we do not overindulge in our daily behaviors. We are constantly bombarded by advertisements, Instagram, and other social media influencers showing us their fabulous lifestyles, and a whole host of other influences that try to direct our attention toward what's new, sexy, innovative, and why we just must have "it," whatever it may be. New technologies, new phones with new features, luxury cars, bespoke watches and jewelry, high fashion, the latest food trend — no matter where we turn or look, something exciting and new is always being revealed as a must have — just got to have — opportunity for us.

It can be incredibly challenging to live within our hyper-materialistic, consumer-driven society. This said, it becomes more paramount for us to fully evaluate our individual and collective goals regarding what we want to accomplish with our lives and between the bookends. Taking the time to thoughtfully think through the virtues, values, and principles that pertain to your pursuit of prosperity and idea of success is essential in how you ultimately define your goals and how you choose to spend your time.

No one can tell you what the right or wrong choices are in the exercise. The decision to have material items is not necessarily bad. And the choice to not acquire material items is not necessarily good. The pursuit of prosperity is not exclusively about accumulating more material possessions. Placing greater priority on attaining more "stuff" as opposed to or in detriment of pursuing other virtues can lead to a decline in prosperity. When we finally stop celebrating and glorifying materialism as success, we can allow our human intellect to rise far above the low bar that it serves in shaping a life well spent.

When we choose to live a life grounded by virtues and allow principles of pragmatism to direct our daily choices and decisions, we can be better prepared and have greater confidence that we are tapping into our inner wisdom to live courageously, and with a sense of moderation, justice, and balance with nature.

When I was younger, particularly during my college years, I often chastised older generations, including the Baby Boomer generation, for their incessant thirst for material possessions. As a Gen-Xer, it felt overwhelming, ridiculous, and even disgusting that the Baby Boomers had accumulated material wealth and driven a global economy focused on fast-and-cheap consumerism and materialistic pursuits above the goals of the common good. While harsh and overgeneralized, I've known too many Baby Boomers who like to show off their new car and their antique car, their big box home and their vacation home, their curated collection of housewares, and their newest home appliance.

Voracious hoarders, Baby Boomers have driven unsustainable consumption in the pursuit of prosperity shaped by their values and beliefs. I've known many a Baby Boomer who puts more attention on shining their car or tidying up their home than they do on spending time with their children or grandchildren. When I experience this, I am saddened, for them and for their loved ones. When we devalue time, take it for granted, or altogether don't consider it as a crucial measure of our virtue, we lose the reins on being masters of our minds, and subsequently, masters of our future prosperity.

For decades, the capacity of the Earth to meet the resource needs of global citizens has been constrained. The idea of this and the reality of how it has played out over the past many decades can feel infuriating. Why must we want for and have so many material possessions? Is the accumulation of more stuff indicative of success and a life well spent? Is it necessary to keep up with the proverbial Joneses or Kardashians? Can joy and happiness be found outside of a consumer-focused economy? Can prosperity be found in something as simple, innocent, and elegant as sitting and watching the serenity of a lake on a beautiful fall day? I think we all know the answer to that question. The container ships have long left the port, and the trains have long left the station. Our consumer-focused society is overflowing with stuff to fill our homes, to sit in our driveways, and to fill all those empty voids within our troubled souls.

As I worked into my 30s and 40s, and as I made the transition from living in an apartment to owning my first home, and then a second home as my family grew, I had to contend with the reality of becoming my parent. I had to take a long, hard look in the mirror and ask myself if I, too, had fallen victim to the consumer lifestyle that I once thought excessive. The choice to be a thoughtful consumer was, as it has always been, within my own self-control. We all have the capacity to make wise economic, social, and environmental decisions within our daily lives — for our families, our communities, and for the betterment of society.

The operating system and code that guide our decisions are what we need to evaluate, test, and modify. I no longer completely blame or disparage Baby Boomers for all the challenges of the world, although they are an easy target. Instead of blame, it feels more productive to forgive and to focus on educating the next generation and creating awareness amongst all people regarding how we can adopt virtues, values, and principles that better guide who we are, who we want to be, and where we are going.

Getting back to some core basic tenets of civility, trust, accountability, and pragmatism can and will lead us down a more sustainable path. Although the train has left the station, we can certainly make corrections with regard to where it's going, what it's carrying, and what impact it will have now and into the future. As I've told hundreds of students whom I have had the pleasure of teaching, placing blame or shame on any one individual or consumer group is not only wrong, but also an unproductive exercise in futility.

We need to focus on and modify the systems and the market failures that have occurred within our society and economy. Cutting across generations will require time, thoughtful reflection and introspection, and courageous will amongst a new generational planet-pragmatic leader. We cannot afford to wait for the perfect solution, the best technology, or ubiquitous buying from every single individual and every single constituency in the world. The pursuit of pragmatism for a more prosperous future requires that we take practical, clear-minded steps that are mindful of the broader sustainability challenges associated with all our daily systems-thinking and systems-based decisions. Planet pragmatism is an approach to enhance our critical thinking in the hearing now as we bring forth ancient wisdom and optimize our collective intellect toward enhancing our quality of life and pursuing greater prosperity for people now and into the future.

Getting back to and practicing the ancient virtues of temperance and moderation does not mean that prosperity must suffer. The following Profile in Pragmatism featuring "farmpreneur" Alex Fasulo highlights how she has exercised her agency to manifest greater prosperity by living with the land, not against it. For the better part of the past three years, Alex Fasulo has been "freelancing her way to freedom," as the proprietor of The House of Green in Upstate, New York, a microfarm that is creating thriving sustainable businesses while serving the local community.

Profiles in Pragmatism: A Question and Answer with Alex Fasulo, The House of Green

Meet Alex Fasulo, a "farmpreneur" and founder of The House of Green

Alex Fasulo is a prolific creator and communicator. She is the epitome of an influencer. But more than this, Alex enlightens, entertains, educates, and empowers others through her positive energy and pragmatic sense of personal growth and development. Alex is in her element when she is guiding others to discover and develop their purpose and talent so that it enriches their life and the lives of those around them.

In 2015, Alex began her career working for a corporation but quickly discovered that the corporate culture and environment were restrictive, repressive, and relentlessly demoralizing. As she put it, "I was tired of seeing friends of mine 'spiritually die' in the corporate environment." She took abrupt action to leave the corporate world behind, and without a concrete plan in place, she became a freelancer with Fiverr. This leap of faith, guided by her will and penchant for independence, was met with profound success. Within 6 months, Alex achieved $150,000 in service revenue, affirming her belief that there are alternative ways to construct one's career and life. Alex discovered freelancing provided a living wage, but, more importantly, it led her to reveal her truth as a budding entrepreneur.

Alex's story was picked up by mainstream media, as she was featured on CNBC. Shortly thereafter, Alex wrote and published her first ebook, *"How I Made 6 Figures in 6 Months on Fiverr."* As Alex continued to freelance with Fiverr, she also launched her brand, "The Freelance Fairy," and began to garner significant attention on social media. An author, freelancer, entrepreneur, and media personality — Alex's self-exploration manifested itself into a thriving brand and freelancing career.

Alex's most recent endeavor, "The House of Green[117]," builds upon her formative freelance experiences. The House of Green is an experiment that fuses the rich potential of aeroponic farming with a unique events space, and the practical deployment and demonstration of technology that preserves and protects the natural environment. In 2023, I caught up with Alex for a 1:1 conversation discussing her impressive journey as a sustainable entrepreneur.

What motivates your strong sense of purpose, particularly as a freelancer and entrepreneur?
I discovered that the corporate lifestyle was not for me. I am an artistic and creative personality, and by quitting my full-time job, I gave myself the space and freedom to discover my strengths, but also my truth as an entrepreneur. I began my freelance journey on Fiverr, before the idea of freelancing was even a thing. Early on in my journey, I also saw how friends of mine were "spiritually dying" at their jobs. This reinforced my immersion into freelancing and entrepreneurial discovery. I like being part of something, and this shift in work and life made me feel as if I were part of a movement. At first, freelancing was a means to an end, and an opportunity to earn some extra money and pay bills. But I quickly learned that it led to a deeper truth about me, as an entrepreneur. It was then that I invested more time and energy in this journey, and the rewards soon followed.

I have not one, but several "whys" on what motivates me and drives my energy and enthusiasm toward entrepreneurship. For example, I'm convinced that we [society] have developed enough [good] technology that we can use in a way to not only coexist with nature, but to revive it. If you elevate above what's happening in society today, human-induced destruction of ecosystems, and so much attention on pollution, waste, climate, and so much more, there is a lot of fear and there is a lot of talk. I see a lot of people talking about these problems, but not enough action. I want to lead by example, do something about specific problems, and help people discover that they can do the same. I have no farming experience. If I can jump in and do something that improves how we can productively restore the Earth and ourselves, then others can too. This is how something gets started, through small steps that build momentum, then a movement, and then a transformation.

Reading a book, *Start Small, Start Thinking Small* — a genius book of how everyone has become paralyzed by big picture talk. Everyone becomes paralyzed by global-scale issues. If everyone can start by doing small things, it will help with the bigger picture. Open one little service if you work on jewelry blogs, even if it's one hour a week. There's opportunity and success in that. The author goes into tinier ways in which people are inventing ways to address issues important to them. Not fighting about the planet warming. There is no discussion, really. You must do what you can.

Tell us some more about your background.

I grew up on a farm, riding horses, swimming in creeks, and caring for animals. I love animals. I also love people and always knew that I wanted to somehow help others, particularly in how they could create and live a life of meaning, value, and impact. That penchant for helping people find side hustles and break out of what they believe they are supposed to do with their life has been an underlying drive for how my career has evolved. That love of helping others and caring for the environment and all living things is what has gotten me to this point in my journey. I like to tell others exploring their career path to not be too rigid about boundary conditions.

What is a "Farmpreneur," and please share with us more about your new endeavor, The House of Green

A farmpreneur encompasses taking some land and discovering ways to make money by being creative. Farmpreneurship requires thinking outside the box and monetizing a farm in non-traditional ways. Given how unsustainable the food supply chain has become, we must localize food networks again. The use of grow-towers is a simple and inexpensive way that land acreage can be optimized for healthy, local, high-quality, and affordable food production. Further, simple solutions like food towers enable the farmpreneur to optimize the remaining farm acreage for other income-producing and shared community and economic opportunities.

Farmpreneurship supports enriching the diversity of economic potential while being a steward of the land, offering a more sustainable and robust opportunity for economic security and resilience. Essentially, there are multiple value streams on the land. The creative side of farmpreneurship is in figuring this all out.

Before The House of Green, I had fears about farming. I had, and still get, impostor syndrome, even though I've experienced success as an author, freelancer, and entrepreneur. But sometimes you must get out of your own way and take a risk. I was in Upstate New York and saw some acreage of land for sale, and a sense of what's next overcame me. There was something about the land, the moment, and the emergence of where I would go next that was enough for me to take the leap into the next stage of my entrepreneurial journey. It can be scary at first, and people often ask me about my plan. The reality is that nobody has the plan worked out in one day. Sometimes you need to make the decision, take the step, and move forward. The second step will reveal itself.

So, I purchased seven acres in Upstate, New York. I'm committed to having the farm off-grid to minimize the environmental impact. The farm and the House of Green were established to help others discover how they can replicate an off-grid agriculture business model. There are so many businesses that appear to be simple. For example, tiny home builders. But then you learn that the entrepreneurs have had 20-or-more-year careers as electricians or homebuilders, and you realize, of course, it's simple for them. I'm really trying to help others learn by doing, and in the process, show them not to be fearful of things that they don't know. I want entrepreneurship, and farmpreneurship specifically, to be accessible for others to be a part of.

I also love folklore, including fairies. That was an inspiration for the Freelance Fairy, and I'm now extending that brand into the farm. I envision fairy-style huts with a mini village or ecosystem so that people can rent a fairy hut for personal use. I'm currently working with local Amish people on some of the site development, most recently employing them to build a driveway into the property. I will also commission the Amish to construct a fairy-style hut. While this may sound whimsical, it's actually practical because Upstate New York gets a lot of wet, heavy snow. The snow load is a critical factor in building, and the steeper pitched roofs of the fairy huts will serve as both an on-brand, aesthetically unique feature of the farm and a functional benefit during snow season.

How do you maintain such a positive attitude and "freelance flow?"
There's a law in physics that states energy can neither be created nor destroyed. I believe this to be true also when it comes to the energy that feeds or depletes people, ideas, and relationships. I focus my energy positively. I'm a huge fan of spreading my attention and energy around, so to speak, so that you don't "block the one thing." Learning from my prior experiences, I now tell others, Don't just quit your job without a plan. Better yet, develop your side-hustle while working your full-time job. I subscribe to the universal Law of indifference — that in the absence of any evidence to the contrary, you should distribute your beliefs equally among all possible outcomes of consideration. In essence, if you become indifferent to success, it tends to happen. In my experience, if you show the Universe that you are cool with what can come your way, then it will manifest and come your way.

Are there any additional things you would like to share?
Regarding The House of Green, and my farmpreneurship journey, I'm open to any ideas for collaboration. As I mention on my social media channel posts, I welcome the chance to speak with others, particularly if they have or share futuristic farming interests. I also welcome anyone who wants to bring their 'farmpreneur' ideas, technologies, brands or businesses to Northern Saratoga County, New York, and collaborate on the land. This is what the House of Green is all about. I'm open to however this takes shape and what happens next.

I'd also like to let anyone with an interest in farmpreneurship know that there are many resources out there to help get you started. In the US, there are so many programs and loans available for buying land. So, if you're interested in taking a leap toward owning land, but think you cannot afford land, that is not true. Check out the U.S. Department of Agriculture (USDA) website; they have so many programs for helping people get started in farming, especially women and minorities. The USDA has incentives if you buy real estate and farms. Essentially, the government would prefer that citizens and the market economy grow and provide food, as opposed to the government doing so. The USDA is a great resource to get started and to access incentives. I also encourage others with interest in farmpreneurship to check out Farm Credit[118], a customer-owned cooperative comprised of a national network of farmers and agriculture-based businesses that support finance and lending for real estate, agriculture production, and in support of rural communities.

Alex Fasulo's book, "Freelance Your Way to Freedom," is available on Amazon[119], Barnes & Noble, and Alex's website[120]. A writer by trade, Alex is always filling her notebooks with new ideas and musings on life as a freelancer. She plans to write another book which captures her experiences with The House of Green and as a farmpreneur. She is looking forward to having others discover just how easy and accessible being a farmpreneur can be.

Pillar IV. The pursuit of planet pragmatism and attainment of prosperity do not have to be a paradox. Life is not so much about what you have or attain; it's about when and where you jump in, what you choose to give, and how you choose to serve.

Have you ever been asked the question, *"What do you do for a living?"* When asked that question, how did you respond? Did you lead with your job title and career path? Or did you offer a different type of response, perhaps focused more on your values and principles? Often, it is much easier to recite your job title than it is to get into the underlying motivations that guide your life's journey.

In American culture, we place an enormous amount of emphasis on our professional identity. Too often, our professional identities become all-consuming and envelop our entire being. How our family, friends, colleagues, and perfect strangers perceive, treat, and communicate with us is largely influenced by our job titles and the persona we take on and present. But who we are cuts much deeper than a job title. Every single one of us is more than their professional title.

Although titles including nurse, law enforcement officer, doctor, lawyer, president and CEO, teacher, counselor, coach, engineer, project manager, firefighter, social worker, and veterinarian are all impressive, the people behind the title are what give meaning and deliver impact to the work that they do. Who we are, as individuals and as a global community of citizens, is much larger than the sum of all our professional titles.

This is not to say that we should not self-identify with job titles, for they do capture some essence of who we are. Job titles are important to our professional identity and shaping our careers. However, we should not take for granted who we are, what we believe, and what we represent underneath the veil of a professional title. The totality of our being is more consequential to achieving prosperity and will outlast any career path.

Jobs provide us with the opportunity to earn a wage to support our lifestyles. For some people, jobs are a means to an end, a transactional relationship of give-and-take. For others, jobs are more than transactional; they are opportunities for putting practical skills and their unique talent into action. And yet for others, jobs represent a deeper conviction, a call to a higher purpose, or an all-encompassing vocation.

Some people have their jobs and titles but also choose to assign their talent and time toward "extracurricular endeavors" that allow them to exercise their values and principles in other ways they cannot do at their job. Some people volunteer, others serve on non-profit or community-based boards, others run for public office, some work with faith-based organizations, and so on. There is no shortage of opportunities or ways in which people give their time, energy, intellect, and talent, beyond the identity of their professional title.

This gets me back to a previous idea and point regarding how we use our time and have the freedom to operate. Instead of the question, "What do you do for a living?" Imagine if you were asked, "How do you like to use your time?" The answer, I bet, would be completely different. Instead of defaulting to reciting a title and an organization, you likely would provide an answer that speaks to something that brings you joy. And joy is rarely selfish.

The greatest joy is when we feel the joy of others; that is when we know that we somehow served a role in providing comfort, happiness, and well-being to another human. This can be as simple as sharing your time with someone who may be lonely, exchanging stories, and having conversations. Or this can be found in watching the fear shift to confidence and then enthusiasm in your son or daughter when you teach them how to drive. One does not need an advanced degree, micro credential, specialty trade, big bank account, fancy car, or designer clothes to bring joy to others. One only needs to be willing to give of themselves to others, as they already are, to discover and share joy, and have a meaningful impact and life.

So, who are you? What are you contributing to the world, right here and now? More importantly, how do you want to use your time? The answer to this question does not need to be as profound as solving world peace, inventing time travel, or eradicating hunger and disease, although each of these are certainly noble pursuits. The answer is for you to discover. The answer can be found in how you align with and enact strong values, and how you choose to lead a principled life through the act of selfless and joyful service and giving.

Pillar V. Prosperity happens when we willingly choose to act together, with common sense, and toward a common good. We gain individual knowledge by doing; we gain societal wisdom by sharing. We all become more prosperous when we reinforce the positive intentions and actions of each other.

The litany of global social, economic, and environmental challenges before our generation is immense, and outright overwhelming. We cannot wait for the perfect time, technology, or temperament of a leader to act. A better world will not magically arrive to greet us. Prosperity, like freedom, is an active pursuit that cannot be taken for granted. We all must work to achieve a better quality of life. Prosperity is never completely given or granted. We all must be active stewards in shaping our present and future conditions.

Having sound virtues and principles are fundamental to living a rich, joyful, and prosperous life. Pontificating our virtues and principles alone cannot ensure planet prosperity. At some point, we need to act resolutely and swiftly for change to occur. Call it what you want, but planet pragmatism requires that we have a "roll up your sleeves and jump in, all hands-on deck, swim or sink, just do it" attitude.

It is essential that we bring a clear and logical mindset to the challenges before us. And it is critical that we realign our values so that we can discover how to live prosperously within nature's laws and with personal and collective temperance, humility, and respect for all living things.

If we are to make any progress toward planet pragmatism and achieve a better quality of life for today, and create opportunities for a better future, then we must have the courage to act, learn, and lead with common sense, practicality, and results. In doing so, we must be open to actively learning through our experiences, share what we have learned with others, and improve upon our state of knowledge with a newfound understanding and a willingness to continuously pursue greatness for all.

Principled Prosperity through Planet Pragmatism — The Word is Out

However this book hits you, the reality for many in the sustainability space remains the same. Now is not a time for dreadful retreat. Rather, it is a time for unity and resolve. You, me, and we make up the common denominator for manifesting and maneuvering toward a better world.

Guided by *common sense for the common good*, we can leverage the knowledge, resources, and leadership of the world's sustainability community to bring about lasting change, peace, and prosperity informed by planet pragmatism – that is, a shared value for caring for one another and for all life on Earth.

Let's not let things get out of hand more than they are. Speak up, speak out, and be proud to be a pragmatist who understands the necessity for protecting the planet as integral to our pursuit of freedom, peace, and prosperity.

Points on Pragmatism

- *Pillar I. Mind mastery, rather than our mastery over nature, may be one of the most compelling and pragmatic remedies toward creating a more sustainable planet. Wisdom is revealed in the calm quiet, when our minds connect to a higher consciousness. To accomplish this, we must manage the rampant and unproductive thoughts within our minds. Mastery of one's mind is a cornerstone of individual compassion, leading to a more benevolent society and greater prosperity.*
- *Pillar II. Learn objectively from the past and lean intentionally into the future. Achieving prosperity requires us to reframe how we define and determine 'success.' Time is our most precious asset, more so than money or technology. Respecting time and using time wisely is a key pillar of 'Planet Pragmatism.'*
- *Pillar III. Temperance and moderation are ancient philosophies that remain pillars of 'Planet Pragmatism.' Measuring worth in material possessions is false and fleeting. When lived with virtue and in a principled way, life delivers deeper riches. You, me, and we can live prosperously with very little.*
- *Pillar IV. The pursuit of planet pragmatism and attainment of prosperity do not have to be a paradox. Life is not so much about what you have or attain; it's about when and where you jump in, what you choose to give, and how you choose to serve.*
- *Pillar V. Prosperity happens when we willingly choose to act together, with common sense, and toward a common good. We gain individual knowledge by doing; we gain societal wisdom by sharing. We all become more prosperous when we reinforce the positive intentions and actions of each other.*

ACKNOWLEDGEMENTS

With deepest gratitude, admiration, and love, I'd like to thank my family, including my life partner, personal confidant, and true love, Aileen, and our two incredibly talented and wonderful sons, Owen and Neal. You are my joy, my peace, and my inspiration. I'm forever grateful for the laughter and love that we all share. I am truly blessed and the luckiest man to have the opportunity to be a part of your lives. I love each of you dearly.

My sincere thanks go out to Wildebeest Publishing's Laura Thorne and Jess Neiding, two extraordinary humans. This book would not have manifested without their vision, creativity, guidance, and support. I am ecstatic and honored to be a member of Wildebeest Publishing's eclectic and diverse community of Authorpreneurs. When I had the conceptual idea for this book a couple of years ago, I knew that I wanted to work with a smaller independent publisher that understood the writer's journey and who demonstrated *principles of pragmatism* in their own journey through strong character, grit, and fortitude for pursuing prosperity and achieving success. I am so grateful to Laura and Jess and the entire Wildebeest team and community. May this book represent a small piece within the beautiful mosaic that Wildebeest is creating.

Writers are influenced by many things and many people. At least in my case, by everyone that I encounter. I've been blessed to have been influenced by some great people and impacted by some unfortunate souls. Experiences, good or bad, shape who we are, how we think, and our worldview. There are too many people who have contributed to how I process the world to identify and thank. However, I must give a shout-out to those who served a role in the ideation and production of this fourth book. My thanks to some dear friends and colleagues who have supported my professional pursuits and creative work for years: Denny Minano, David Montanaro, Steve Myers, Pamela Norton, Rajiv Ramchandra, Frank Reilly, David Tarino, Stelios Vogiatzis, and Joe Zagrobelny. Each of you put your faith in me, for which I am thankful.

A very special thanks to the Rev. Brian Konkol for his wise counsel, collegiality, and friendship, all of which inspired the opening poem and introduction to this book. Brian, whenever we talk, I always walk away feeling reinvigorated to serve and to continue to be a better human. You are wise beyond your years; you always remind me that we could all benefit from more faith in our lives. My warmest appreciation for Princy Mthombeni and Alexandra Fasulo, who contributed to the profiles in Pragmatism features in this book. Your respective work is awe-inspiring and so hopeful. When I think about your work and leadership, I'm optimistic about our shared future.

I've dedicated this book to all the students that I have had the pleasure of working with for several years now. I hope that I have been a source of knowledge and inspiration and have enriched your lives as much as you have done for me. This book is a celebration of how humanity, when we are at our finest, can learn, lead, and grow together. The act and art of teaching and the profession of education have never been a one-way street. Knowledge is gained through shared experience, but also through continuous testing, evaluation, and transfer of ideas into practice. I'm so thankful to my former students for their curiosity, creativity, compassion, courage, confidence, and clarity that have deepened our class discussions and led to critical thought, the derivation of new ideas, and personal growth and development. Each of you has such a bright future ahead. Please know that I am always available to support your ongoing pursuit of personal growth and prosperity.

Finally, to my extended family, friends, and colleagues. Thank you for your ongoing and unwavering support. You've enriched my life, and I hope that I have reciprocated in kind. I wish everyone peace, love, and principled prosperity! Thank you for being a part of this creative journey.

REFERENCES & ENDNOTES

For more information, please visit the author's website: www.mark-colemaninsights.com

1. Impakter, Accessed March 14, 2024. Coleman, Mark. "A Silent Spring Has Fallen Upon Us, It's Time to Listen so That We Can Emerge with Dignity." https://impakter.com/a-silent-spring-has-fallen-upon-us-its-time-to-listen-so-that-we-can-emerge-with-dignity/
2. Wikipedia, Accessed March 4, 2024. https://en.wikipedia.org/wiki/Airbnb
3. Wikipedia, Accessed March 4, 2024. https://en.wikipedia.org/wiki/Rent_the_Runway
4. Wikipedia, Accessed March 4, 2024. https://en.wikipedia.org/wiki/Uber
5. Wikipedia, "Blockchain." Accessed February 12, 2025. https://en.wikipedia.org/wiki/Blockchain
6. Agile District, LinkedIn. September 30, 2022. "Web 4.0 Explained – A Brief!" https://www.linkedin.com/pulse/web-40-explained-brief-agiledistrict/
7. Wikipedia, "Decentralized Finance." Accessed February 12, 2025. https://en.wikipedia.org/wiki/Decentralized_finance
8. United Nations, https://www.un.org/en/academic-impact/sustainability

9. Aguilar, R. Andres Castaneda. March 26, 2024. "March 2024 global poverty update from the World Bank: first estimates of global poverty until 2022 from survey data." World Bank. https://blogs.worldbank.org/en/opendata/march-2024-global-poverty-update-from-the-world-bank--first-esti
10. Wikipedia, Accessed March 11, 2024. https://en.wikipedia.org/wiki/Seven_generation_sustainability
11. United Nations Sustainable Development Goals, Accessed March 11, 2024. https://www.un.org/sustainabledevelopment/sustainable-development-goals/
12. UN Global Compact. https://unglobalcompact.org/
13. Simon Sinek, https://simonsinek.com/
14. SmartInsights, Accessed March 8, 2024. https://www.smartinsights.com/digital-marketing-strategy/online-value-proposition/start-with-why-creating-a-value-proposition-with-the-golden-circle-model/
15. Yvon Chouinard, https://en.wikipedia.org/wiki/Yvon_Chouinard
16. Hamdi Ulukaya, https://en.wikipedia.org/wiki/Hamdi_Ulukaya#:~:text=Hamdi%20Ulukaya%20(born%2026%20October,yogurt%20brand%20in%20the%20US.
17. Sara Blakely, https://en.wikipedia.org/wiki/Sara_Blakely
18. Jessica Alba, https://en.wikipedia.org/wiki/Jessica_Alba
19. Ecology Prime, https://ecologyprime.com/the-ep-mission/
20. Clover Hogan, LinkedIn, accessed 4/4/24, https://www.linkedin.com/posts/cloverhogan_cop29-justice-equity-activity-7181600323579318272-8c-F?utm_source=share&utm_medium=member_desktop
21. Deloitte. "The Challenge of Double Materiality." https://www2.deloitte.com/cn/en/pages/hot-topics/topics/climate-and-sustainability/dcca/thought-leadership/the-challenge-of-double-materiality.html
22. World Meteorological Organization (WMO). "Greenhouse gas concentrations surge again to new record in 2023." October 28, 2024. https://wmo.int/media/news/greenhouse-gas-concentrations-surge-again-new-record-2023

23. International Energy Agency (IEA). "Growth in global oil demand is set to slow significantly by 2028." June 14, 2023. https://www.iea.org/news/growth-in-global-oil-demand-is-set-to-slow-significantly-by-2028
24. National Geographic. https://www.nationalgeographic.com/environment/article/dubai-ecological-footprint-sustainable-urban-city
25. Page, Tom; and Max Burnell. CNN. April 25, 2019. "$13.6B record-breaking solar park rises from Dubai desert." https://www.cnn.com/style/article/mbr-solar-park-dubai-desert-intl/index.html
26. NEOM. https://www.neom.com/en-us
27. Wikipedia. Red Sea Global. Accessed October 28, 2024. https://en.wikipedia.org/wiki/Red_Sea_Global#:~:text=6%20External%20links-,History,%2C%20Qiddiya%2C%20and%20Diriyah%20Company.
28. Wikipedia. List of Saudi Vision 2030 Projects. Accessed October 28, 2024. https://en.wikipedia.org/wiki/List_of_Saudi_Vision_2030_projects
29. Toyota Woven City. https://www.woven-city.global/about/
30. International Energy Agency (IEA). July 19, 2024. "Global electricity demand set to rise strongly this year and next, reflecting its expanding role in energy systems around the world." https://www.iea.org/news/global-electricity-demand-set-to-rise-strongly-this-year-and-next-reflecting-its-expanding-role-in-energy-systems-around-the-world
31. Orf, Darren. September 5, 2024. Popular Mechanics. "Whoopsie, SpaceX Blew Up Two Rockets and Punched a Massive Hole in One of Earth's Layers." https://www.popularmechanics.com/space/rockets/a62047078/starship-explosion-ionosphere/
32. Wikipedia. "Love Canal" https://en.wikipedia.org/wiki/Love_Canal
33. Wikipedia. "A Civil Action" https://en.wikipedia.org/wiki/A_Civil_Action_(film)

34. Wikipedia. "Pollution of the Hudson River" https://en.wikipedia.org/wiki/Pollution_of_the_Hudson_River
35. For additional reference, see https://www.markcolemaninsights.com/post/the-new-accounting-real-and-lasting-sustainability-happens-in-the-moment
36. Wikipedia. Accessed August 7, 2024. "East Palestine, Ohio, train derailment." https://en.wikipedia.org/wiki/East_Palestine,_Ohio,_train_derailment
37. Wikipedia. "Peter Senge." Accessed June 10, 2024. https://en.wikipedia.org/wiki/Peter_Senge
38. Osterwalder, Alex. August 29, 2012. Strategyzer. "Achieve product-market fit with our brand-new Value Proposition Canvas." https://www.strategyzer.com/library/achieve-product-market-fit-with-our-brand-new-value-proposition-designer-canvas
39. Wikipedia. "Tesla, Inc." Accessed February 25, 2025. https://en.wikipedia.org/wiki/Tesla,_Inc.
40. Shahan, Zachary. CleanTechnica. May 5, 2024. "Tesla Still Sells More EVs In USA Than Ford, Chevrolet, Hyundai, Kia, Audi, BMW & Toyota Combined." https://cleantechnica.com/2024/05/04/tesla-still-sells-more-evs-in-usa-than-ford-chevrolet-hyundai-kia-audi-bmw-chevrolet-toyota-combined/#:~:text=6'-,Tesla%20Still%20Sells%20More%20EVs%20In%20USA%20Than%20Ford%2C%20Chevrolet,BMW%2C%20Chevrolet%2C%20%26%20Toyota%20Combined&text=Actually%2C%20wait%2C%20Tesla%20still%20sells,sales%20in%20the%20first%20quarter.
41. "Tesla, Inc. ESG Score." S&P Global. Tesla ESG Score as of January 27, 2025. https://www.spglobal.com/esg/scores/results?cid=4574287
42. Microsoft. https://news.microsoft.com/source/features/sustainability/microsoft-builds-first-datacenters-with-wood-to-slash-carbon-emissions/

43. Patagonia. https://www.patagonia.com/our-responsibility-programs.html
44. Wikipedia. "Wal-Mart." https://en.wikipedia.org/wiki/Walmart
45. Herrel, Katie. December 4, 2024. Backpacker. "Walmart Announces Green Labeling" https://www.backpacker.com/news-and-events/walmart-announces-green-labeling/
46. Walmart. Walmart Recycling Playbook. https://www.walmartsustainabilityhub.com/content/dam/walmart-sustainability-hub/documents/project-gigaton/packaging/walmart-recycling-playbook.pdf
47. IKEA. "The story of IKEA flatpacks." https://www.ikea.com/ph/en/this-is-ikea/about-us/the-story-of-ikea-flatpacks-puba710ccb0/
48. Wikipedia. "IKEA." Accessed February 27, 2025. https://en.wikipedia.org/wiki/IKEA#cite_note-115
49. Wikipedia. "The Natural Step." Accessed February 27, 2025. https://en.wikipedia.org/wiki/The_Natural_Step
50. IKEA. 2021 Sustainability Report. https://www.ikea.com/global/en/images/ikea_sustainability_report_fy21_4d253ede75.pdf
51. United Nations Environment Programme, "Five ways countries can adapt to the climate crisis." October 12, 2022. https://www.unep.org/news-and-stories/story/5-ways-countries-can-adapt-climate-crisis
52. Reuters, March 14, 2024. 'Climate adaptation investor roadmap points way through global warming.' https://www.reuters.com/sustainability/sustainable-finance-reporting/climate-adaptation-investor-roadmap-points-way-through-global-warming-2024-03-14/#:~:text=Climate%20adaptation%20investor%20roadmap%20points%20way%20through%20global%20warming,-Reuters&text=March%2014%20(Reuters)%20%2D%20Faster,in%20a%20report%20on%20Thursday.
53. U.S. Environmental Protection Agency, "Climate Change Indicators: Wildfires." https://www.epa.gov/climate-indicators/climate-change-indicators-wildfires#:~:text=Although%20

wildfires%20occur%20naturally%20and,wildfire%20frequency%2C%20and%20burned%20area.

54. International Energy Agency (IEA). "The Oil and Gas Industry in Net Zero Transitions: World Energy Outlook Special Report." 2023. https://www.iea.org/reports/the-oil-and-gas-industry-in-net-zero-transitions
55. OPEC 2024 World Oil Outlook 2050. https://publications.opec.org/woo/Home
56. United Nations. "United Nations Decade on Restoration." https://www.decadeonrestoration.org/what-ecosystem-restoration#:~:text=Ecosystem%20restoration%20means%20assisting%20in,ecosystems%20that%20are%20still%20intact.
57. Washington Post. January 14, 2025. https://www.washingtonpost.com/weather/2025/01/14/los-angeles-fires-california-palisades-eaton-updates/
58. Case Study on the California Blackouts: https://education.nationalgeographic.org/resource/case-study-california-blackouts/
59. Grenier, Letitia, with Jeffrey Mount, Sarah Bardeen, Ellen Hanak, Bradley Franklin, Kyle Greenspan, Spencer Cole, Brian Gray, and Gokce Sencan. "Priorities for California's Water: Are We Ready for Climate Change?," November 2024. https://www.ppic.org/publication/priorities-for-californias-water/
60. CNN. October 8, 2024. "Hurricane Milton's magnitude brings veteran meteorologist to tears" https://www.cnn.com/2024/10/08/weather/video/hurricane-milton-meteorologist-tears-up-tv-ovn-ldn-digvid
61. World Wildlife Fund, "Catastrophic 73% decline in the average size of global wildlife populations in just 50 years reveals a 'system in peril'." Accessed October 11, 2024. https://www.worldwildlife.org/press-releases/catastrophic-73-decline-in-the-average-size-of-global-wildlife-populations-in-just-50-years-reveals-a-system-in-peril
62. World Wildlife Fund. October 9, 2024. "Catastrophic 73%

decline in the average size of global wildlife populations in just 50 years reveals a 'system in peril'" https://www.worldwildlife.org/press-releases/catastrophic-73-decline-in-the-average-size-of-global-wildlife-populations-in-just-50-years-reveals-a-system-in-peril

63. Ecosystem Services. National Wildlife Federation. https://www.nwf.org/Educational-Resources/Wildlife-Guide/Understanding-Conservation/Ecosystem-Services#:~:text=A%20provisioning%20service%20is%20any,other%20materials%2C%20and%20medicinal%20benefits.
64. Wartsila, "The AI gender gap: living in a world designed by men." Accessed April 2, 2024. https://www.wartsila.com/insights/article/the-ai-gender-gap-living-in-a-world-designed-by-men
65. Burns, Robert. "To a Mouse." Poetry Foundation. https://www.poetryfoundation.org/poems/43816/to-a-mouse-56d222ab36e33; see also, "To a Mouse." https://en.wikipedia.org/wiki/To_a_Mouse
66. Wikipedia. "Robert Burns." Accessed April 15, 2024. https://en.wikipedia.org/wiki/Robert_Burns
67. Wikipedia. "PDCA." Accessed April 15, 2024. https://en.wikipedia.org/wiki/PDCA
68. "Logic Model Readings." Center on Philanthropy and Civil Societ. Stanford University. Accessed April 15, 2024. https://pacscenter.stanford.edu/wp-content/uploads/2015/07/Logic-Model-Readings.pdf; see also, "An expanded simple logic model." University of Wisconsin-Madison. Accessed April 15, 2024. https://logicmodel.extension.wisc.edu/introduction-overview/section-1-what-is-a-logic-model/1-7-an-expanded-simple-logic-model/
69. United Nations Development Group, "Theory of Change." Accessed April 15, 2024. https://unsdg.un.org/sites/default/files/UNDG-UNDAF-Companion-Pieces-7-Theory-of-Change.pdf
70. Center for Theory of Change. "What is Theory of Change?"

Accessed April 15, 2024. https://www.theoryofchange.org/what-is-theory-of-change/

71. Wells College Center for Sustainability and the Environment. Accessed April 16, 2024. https://www.wells.edu/wp-content/uploads/center_sustainability/Sustainable_Business_flyer_-_Mark_Coleman_October_4_2022.pdf See also, https://www.youtube.com/watch?v=0lmYlVhNRsk&feature=youtu.be
72. Our World in Data. "Urbanization." Accessed April 17, 2024. https://ourworldindata.org/urbanization#:~:text=By%202050%2C%20close%20to%207,the%20UN's%20medium%20fertility%20scenario.
73. Wikipedia. Clive Humby. Accessed July 29, 2024. https://en.wikipedia.org/wiki/Clive_Humby
74. United Nations Sustainable Development Goals (SDGs), https://sdgs.un.org/goals
75. Alvin Toffler, The Third Wave https://www.pwc.com/gx/en/government-public-services/assets/five-megatrends-implications.pdf
76. Kelly, Leanne M., and Maya Cordeiro. "Three principles of pragmatism for research on organizational processes." May-August 2020. Methodological Innovations. https://journals.sagepub.com/doi/pdf/10.1177/2059799120937242#:~:text=These%20principles%20are%20(1)%20an,inquiry%20as%20an%20experiential%20process
77. Nestle, Marion. "Least credible food industry ad of the week: JBS and climate change." Food Politics. April 26, 2021. https://www.foodpolitics.com/2021/04/least-credible-food-industry-ad-of-the-week-jbs-and-climate-change/
78. Dutkiewicz, Jan. "Why New York is suing the world's biggest meat company." March 8, 2024. https://www.vox.com/future-perfect/2024/3/8/24093774/big-meat-jbs-lawsuit-greenwashing-climate-new-york
79. United States Department of Agriculture (USDA), "Ecosystem Services." https://www.climatehubs.usda.gov/

ecosystem-services#:~:text=Ecosystem%20services%20are%20 the%20direct,support%20and%20sustain%20human%20 livelihoods.

80. Kesherim, Ruben. October 5, 2023. "Average Human Attention Span (By Age, Gender & Race). Supportive Care
81. Bruck, Hannah. July 16, 2024. Texas Public Policy Foundation. "The Energy Cost of Social Media." https://www.texaspolicy.com/the-energy-cost-of-social-media/
82. Vreeswijk, Simon. Shift. November 27, 2023. "The Carbon FootprintoftheInternet:HowYourDataUsageEmitsCO2."https://shift.com/blog/news/the-carbon-footprint-of-the-internet/
83. International Energy Agency (IEA), World Energy Investment, 2023. https://www.iea.org/reports/world-energy-investment-2023/overview-and-key-findings#abstract
84. U.S. Energy Information Administration (EIA), https://www.eia.gov/tools/faqs/faq.php?id=77&t=3
85. U.S. Department of Energy Lawrence Berkeley National Laboratory, https://emp.lbl.gov/queues
86. U.S. Department of Energy, "DOE Releases First-Ever Roadmap to Accelerate Connecting More Clean Energy Projects to the Nation's Electric Grid." April 17, 2024. https://www.energy.gov/articles/doe-releases-first-ever-roadmap-accelerate-connecting-more-clean-energy-projects-nations?utm_campaign=&utm_content=1713635101&utm_medium=U.S.+Department+of+Energy+%28DOE%29&utm_source=linkedin
87. Wikipedia. "Nuclear Warfare." Accessed April 29, 2024. https://en.wikipedia.org/wiki/Nuclear_warfare#:~:text=During%20 the%20final%20stages%20of,have%20been%20used%20in%20 combat.
88. U.S. Department of Energy. 5 Incredible Ways Nuclear Powers Our Lives. October 12, 2018. https://www.energy.gov/ne/articles/5-incredible-ways-nuclear-powers-our-lives
89. Wikipedia. Nuclear power by country. Accessed April 30, 2024. https://en.wikipedia.org/wiki/Nuclear_power_by_country

90. U.S. Department of Energy. "Advanced Small Modular Reactors." https://www.energy.gov/ne/advanced-small-modular-reactors-smrs
91. Fusion Industry Association. https://www.fusionindustryassociation.org/
92. The meeting, "Reliable Decarbonization in the Northeast – Dialogues, Policies and Innovation," took place on Thursday, June 9, 2022, and was held at the Normandy Farm, Blue Bell, Montgomery County, Pennsylvania. Agenda available at: https://www.eckertseamans.com/app/uploads/Agenda-for-Wind-Conference.pdf
93. For additional background on Joseph Dominguez, President and CEO of Constellation, see https://www.constellationenergy.com/our-company/leadership/executive-profiles/joseph-dominguez.html
94. Source: https://www.constellationenergy.com/our-company/leadership/executive-profiles/joseph-dominguez.html
95. Lawrence Livermore National Laboratory, U.S. Department of Energy, https://gs.llnl.gov/energy-homeland-security/energy-security/energy-flow-charts
96. Lawrence Livermore National Laboratory, U.S. Department of Energy, https://flowcharts.llnl.gov/
97. International Energy Agency (IEA). August 29, 2023. "Global power sector saved fuel costs of USD 520 billion last year thanks to renewables, says new IRENA report." https://www.irena.org/News/pressreleases/2023/Aug/Renewables-Competitiveness-Accelerates-Despite-Cost-Inflation
98. Wikipedia. Laws of Thermodynamics. Accessed February 11, 2025. https://en.wikipedia.org/wiki/Laws_of_thermodynamics
99. Wikipedia. 'Cinereus shrew.' Accessed October 15, 2024. https://en.wikipedia.org/wiki/Cinereus_shrew
100. IMDb. "Little Shop of Horrors, quotes." https://www.imdb.com/title/tt0091419/quotes/
101. Wikipedia. Accessed October 15, 2024. "No such thing

as a stupid question." https://en.wikipedia.org/wiki/No_such_thing_as_a_stupid_question

102. Rouch, Maeghan, and Aaron Denman, Peter Hanbury, Paul Reno, and Ellyn Gray. Bain & Company. "Utilities Must Reinvent Themselves to Harness the AI-Driven Data Center Boom." October 10, 2024. https://www.bain.com/insights/utilities-must-reinvent-themselves-to-harness-the-ai-driven-data-center-boom/
103. Reuters. September 21, 2024. "Microsoft deal propels Three Mile Island restart, with key permits still needed." https://www.reuters.com/markets/deals/constellation-inks-power-supply-deal-with-microsoft-2024-09-20/
104. Meta. August 26, 2024. "New Geothermal Energy Project to Support Our Data Centers." https://about.fb.com/news/2024/08/new-geothermal-energy-project-to-support-our-data-centers/
105. Verma, Pranshu, and Shelly Tan. "A bottle of water per email: the hidden environmental costs of using AI chatbots." The Washington Post. September 18, 2024. https://www.washingtonpost.com/technology/2024/09/18/energy-ai-use-electricity-water-data-centers/
106. BBC. "How did oil come to run our world?" https://www.bbc.co.uk/teach/articles/zn6gnrd
107. Wikipedia. Accessed February 20, 2025. "1973 Oil Crisis." https://en.wikipedia.org/wiki/1973_oil_crisis
108. Dang, Sheila, and Seher Dareen. January 27, 2025. "Big Oil in no rush to 'drill baby drill' this year despite Trump agenda." Reuters. https://www.reuters.com/business/energy/big-oil-no-rush-drill-baby-drill-this-year-despite-trump-agenda-2025-01-27/
109. The Guardian. "All the executive orders Trump has signed so far." February 19, 2025. https://www.theguardian.com/us-news/2025/jan/29/donald-trump-executive-orders-signed-list
110. Will, Madeline. September 5, 2023. "U.S. Teachers Lag Behind Global Peers in Teaching About Sustainability. Here's Why."

EdWeek, Science. https://www.edweek.org/teaching-learning/u-s-teachers-lag-behind-global-peers-in-teaching-about-sustainability-heres-why/2023/09

111. Will, Madeline. September 5, 2023. “U.S. Teachers Lag Behind Global Peers in Teaching About Sustainability. Here’s Why.” EdWeek, Science.
112. Jennifer Phillips Blog, “Origin Story Be the Change.” Accessed April 16, 2024. https://jenniferlphillips.com/blog/2021/2/24/origin-story-be-the-change#:~:text=Be%20the%20change%20you%20wish,We%20but%20mirror%20the%20world.
113. Re:CREATe https://recreateindia.org/vision-mission/
114. Re:CREATe https://recreateindia.org/what-we-do/
115. Rematec.https://www.rematec.com/news/industry-players-and-markets/recreate-wins-best-reman-ambassador-award
116. Wikipedia. “Marcus Aurelius.” Accessed October 2, 2024. https://en.wikipedia.org/wiki/Marcus_Aurelius
117. Alex Fasulo, The House of Green, https://www.alexfasulo.com/house-of-green
118. Farm Credit, https://farmcredit.com/
119. Alex Fasulo, “Freelance Your Way to Freedom.” Amazon. https://www.amazon.com/Freelance-Your-Way-Freedom-Corporate/dp/1119893232/ref=sr_1_1?crid=1RC6RI7RPYB5Z&keywords=freelance%20your%20way%20to%20freedom&qid=1659972758&sprefix=freelance%20your%20way%20%2Caps%2C99&sr=8-1
120. Alex Fasulo website: https://www.alexfasulo.com/freelanceyourwaytofreedom

ABOUT THE AUTHOR

Mr. Coleman is a Strategic Sustainable Enterprise Executive with more than 20 years' experience leading the development and implementation of strategic initiatives that leverage resources, research, and expertise to support local, regional, and national energy and economic development. He has worked for and with leading academic, applied research, engineering and management consulting, advanced manufacturing, and government organizations at the nexus of economic development, social need, and sustainability impact.

As an award-winning author and a recognized voice, strategist, business advisor, entrepreneur, and consultant on sustainable enterprise and the convergence of energy, technology, environmental stewardship, and innovation, he has advised hundreds of organizations in the areas of sustainability, risk, innovation, operational effectiveness, change management, and business strategy and economic development.

Currently, Mr. Coleman serves as the Director, Advanced Energy Advisory and Innovation for TRC Companies, a global leader in energy and environmental engineering services and infrastructure solutions. At TRC, Mark works at the nexus of energy innovation and the emergent sustainable economy, marked by solutions which are decarbonized, digital, decentralized, and which focus on social and environmental justice and equity at their foundation.

Committed to regional sustainable economic development, Mark is currently serving a 6-year term as a Trustee to *Cayuga Community College* in Auburn, New York. He also serves as a member of the Board for *Ecology Prime,* an ecology social impact organization which is "creating the world's go-to resource and publishing platform for students, consumers, and businesses for environmental information and global collaboration." Mark is also currently serving as an advisor to the Recreate India Research Foundation (Re:CREATe), a research and advocacy enterprise that aims to catalyze and advance the remanufacturing industry in India.

An entrepreneur in practice, Mark has also served as the President of Convergence Mitigation Management (CMM), a high-impact business intelligence, strategy, and management consultancy providing custom advisory services to entrepreneurs, small and medium-sized businesses, government, applied research, and non-governmental organizations. Through his leadership at CMM, Mark spent 15 years facilitating productive peer-to-peer executive meetings and workshops among thirty of the world's largest corporations.

Mark formerly held leadership responsibilities with Syracuse University, the New York State Energy Research and Development Authority (NYSERDA), advanced manufacturer HARBEC, Inc., and three prominent organizations at Rochester Institute of Technology (RIT) including the Golisano Institute for Sustainability (GIS), the Center for Integrated Manufacturing Studies (CIMS), and the Clean Energy Incubator (CEI) associated with RIT's Venture Creations.

A prolific writer and thought leader on sustainable enterprise and innovation, Mark's third book, *The Dignity Doctrine: Rational Relations in an Irrational World,* was published in August 2020. The book has been described as a *"welcome [and timely] guide to inspire positive change in the relationships in our daily lives, as well as our goals to advance sustainable enterprise." The Dignity Doctrine* was awarded a 2021 Silver Medal by the Axiom Business Books Awards in the Business Ethics Category. In 2014, Mark's second book, the award-winning title *Time to Trust: Mobilizing Humanity for a Sustainable Future,* was published by MotivationalPress. *Time to Trust* was awarded a Silver Medal by the Axiom Business Books Awards in the Business Ethics category, and in 2015 was also an Award-Winning Finalist in the Social Change category of the 7th Annual International Book Awards sponsored by American Book Fest. *Time to Trust* followed his 2012 seminal book release, *The Sustainability Generation: The Politics of Change and Why Personal Accountability is Essential NOW!,* published by SelectBooks. Mr. Coleman's books highlight his perspective on holistic systems-level logic and theory for advancing humanity beyond the status quo toward more integrated models of sustainable development.

Mark has been an active blogger with the Huffington Post, Impakter, and IntelligentHQ and has published numerous articles with leading organizations and journals, including the American Public Works Association (APWA), GreenBiz.com, Environmental Leader, *Triple Bottom Line Magazine,* and *NOVA Holistic Journal.* Mr. Coleman also served as an adjunct instructor at the Whitman School of Management at Syracuse University, where he has taught undergraduate and graduate classes in Sustainable Enterprise and Managing Sustainability.

Mr. Coleman resides in the Finger Lakes region of New York with his wife Aileen and two sons, Owen and Neal.

www.ingramcontent.com/pod-product-compliance
Lightning Source LLC
Chambersburg PA
CBHW062117020426
42335CB00013B/1003

* 9 7 8 1 9 5 8 2 3 3 4 3 6 *